Cap margin

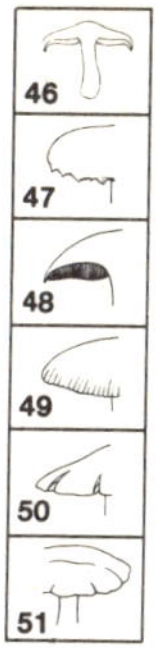

46 Involute, turned inwards

47 Fringed by the remnants of the general or partial veil

48 Overhanging or jutting

49 Radially striate

50 Radially furrowed

51 Wavy or lobed

Tubes

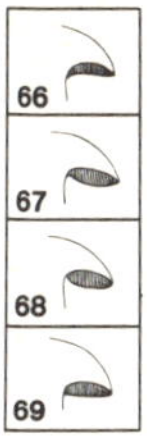

66 Decurrent

67 Decurrent with fillet

68 Free

69 Adnate

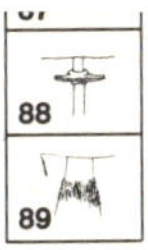

87 Drooping

88 Lifted at the margin

89 False ring, remnant of the cortina

Cortina

90 Stringy or glutinous cortina

Stipe

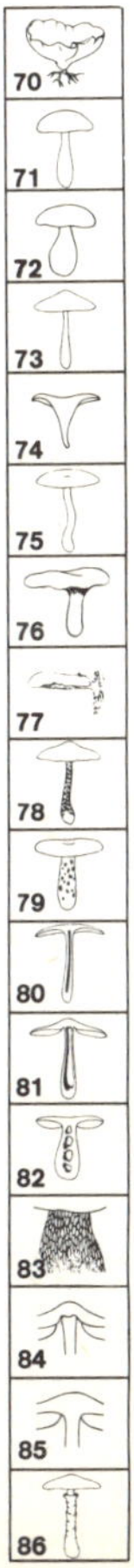

70 Absent, sessile fungus

71 Clavate or club-shaped

72 Pear-shaped

73 Narrowing at the top

74 Widening at the top

75 Wavy or bending

76 Eccentric

77 Lateral

78 Marked by annular areas of different colour

79 Scrobiculated, or pock-marked

80 Fistulous, empty inside like a small tube

81 Tubular, empty like a large tube

82 Cavernous, with one or more hollows

83 Reticulated

84 Heterogeneous (different tissues from those of the cap and easily separated)

85 Homogeneous (same tissue as the cap, can only be broken off)

86 Sheathed to the base of the ring

Foot

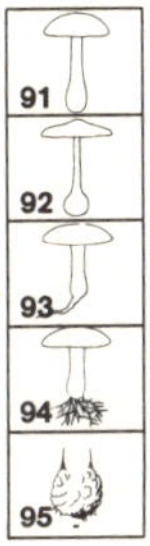

91 Swollen

92 Bulbous

93 Rooting

94 With mycelium fragments

95 Sheathed in mycelium cotton

Gills

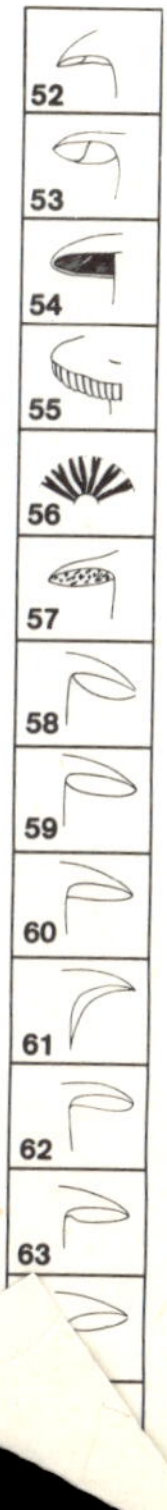

52 Shallow

53 Deep

54 Crowded and, generally, thin

55 Distant and, generally, thick

56 Forked, either with the tip subdivided or joined by transverse veins

57 Marbled, of uneven colour

58 Adnexed to the stipe

59 Free from the stipe

60 Adnate (adhering to the stipe)

61 Decurrent

62 Briefly decurrent by means of a fillet or hook

63 Emarginate

Serrated (on the edge)

Irregularly dented (on the edge)

Volva

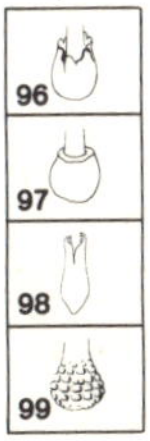

96 Membraneous

97 Truncated, with a smooth brim

98 Sheathing, tall and adhering to the stipe

99 Dissociated, breaking up into scales

Milk

100 A fungus exuding milk on being broken or incised

Mushrooms & Toadstools

a colour field guide

Mushrooms & Toadstools

a colour field guide

U. Nonis

English translation by Lucia Wildt

DAVID & CHARLES
Newton Abbot London

Cover illustration: *Amanita muscaria* Fly Agaric (*Gilbert Garot*)
Colour photographs and pictograms: U. Nonis

British Library Cataloguing in Publication Data

Nonis, U.
Mushrooms & toadstools.
1. Mushrooms—Europe—Identification manuals
I. Title II. Funghi a colpo d'occhio. *English*
589.2'22'094 QK617
ISBN 0-7153-8282-9

Filmset by Keyspools Ltd, Golborne, Lancs
Printed in Italy

CONTENTS

FOREWORD

The 168 species of fungi reproduced and described in this book have been selected from the 1500 which we studied, discovered, photographed and identified during many years of detailed research and experiments, sometimes quite dangerous. We actually tested personally (at our own risk, and occasionally with rather disagreeable consequences) the edibility or otherwise, even the toxicity, of some 1287 species, so that we can absolutely guarantee the accuracy of the information contained in these pages.

A. MORPHOLOGY AND CLASSIFICATION OF THE MACROMYCETES

Latin is the official language of botany and its use allows botanists to understand one another no matter their country of origin. However, many scientific names are derived from Greek, and, in order to simplify the understanding of the text, we shall give the etymology of the scientific terms. The etymology of the names of genera and species appears in the glossary at the end of the book. The words 'latin' and 'greek' are shortened respectively to l. and g.

The science concerned with the study of fungi is called mycology (g. *mukes:* mushroom and *logos:* treatise). Fungi are divided into two wide categories: those which produce spores over complex, macroscopic structures, and those whose spores are borne on the filaments of the mycelium, without any need for such structures, and which are therefore still microscopic. The former are called macromycetes (g. *makros:* large) and the latter micromycetes (g. *mikro:* small). Since the micromycetes are not visible to the naked eye, we will not deal with them in the following pages. One should not forget, however, that these 'moulds' play an important role in our life, either a useful one (as sources of antibiotics, or as fermenting agents for bread, beer, wine, cheese) or a detrimental one (parasites of plants, animals and man; destroyers of both materials and foodstuffs). Our only interest here is with the macromycetes.

While the boletes of our woods and the agarics of our meadows are actually called macromycetes, the parts which we collect and bring back to our kitchens or our laboratories are only the fruit bodies of the fungi, which show above the surface of the earth; the vegetative part remains underneath. This part is made up of a cluster of various filaments or hyphae, which form the mycelium. The filaments are extremely fine (their diameter is usually less than one hundredth of a millimetre) and practically invisible if isolated. The mycelium of the macromycetes produces a fruit body called carpophore (g. *karpos:* fruit and *phero:* bearing). Some of the cells of the carpophores will eventually produce the spores which, once spread, will perpetuate the species. The spore-bearing cells can be arranged either on a regular layer covering part of the surface of the carpophore (called hymenium, g. *humen:* membrane) or as a more or less voluminous mass contained within the carpophore (called gleba, l. *gleba:* glebe, sod).

The characteristics both of the spores and of the cells which produce them are very important to the classification of mushrooms; they form the basis of the division of the macromycetes into two main classes: the Basidiomycetes and the Ascomycetes. The former usually carry four spores on the summit of cells called basidia (g. *basidion:* small foot); while with the Ascomycetes the spores are contained within a cell called ascus (g. *askos:* other), usually eight of them, and they are expelled on maturity. All these spores are of course microscopic since their widest diameter is something like one hundredth of a millimetre. Therefore they can easily be transported by the wind over what are often fantastic distances. If they find a favourable habitat, they will germinate and produce a mycelium which will in its turn originate one or more carpophores, if all the necessary conditions are there.

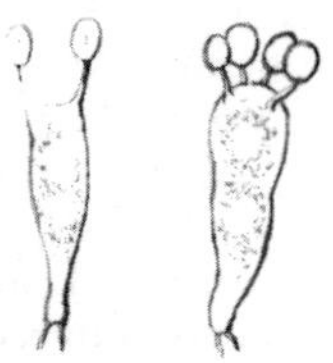

Basidia with spores.

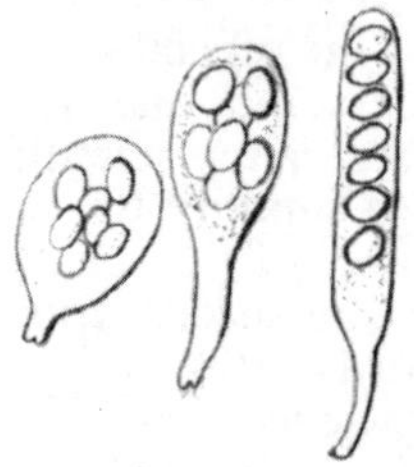

Asci with spores.

At the very first stage of its development, the carpophore, still in its very rough stage, is called the primordium (l. *primordium:* beginning) or sporophore. In some species, the cells which produce spores are at this stage already exposed to the air. Once the carpophore is properly developed, a layer of cells called the hymenium will be formed. Since these spores are borne on bare surfaces, this type of fungus is called gymnocarpous (g. *gumnos:* naked and *karpos:* fruit). Other species bear their spores within the primordium and the carpophore until they are ripe. These fungi are called angiocarpous (g. *angios:* envelope). Finally, a third category has primordia wrapped in one or two veils which they shed as they grow, so that the hymenium is well protected at the beginning, when the spores would otherwise be exposed to the air before being ripe.

Primordium wrapped in the general veil.

The breaking up of the general veil and the formation of warts and volva.

The breaking of the partial veil and the formation of the ring.

These types are called hemiangiocarpes and the veils which surround them are called either the general veil, which wraps up the whole of the primordium, or the partial veil. If the general veil is a membrane, it is shed to allow for the growth of the carpophore and its remains form the volva; this happens in the case of the *Amanita phalloides*. Otherwise, it breaks into many pieces which survive on the growing cap and on the base of the stipe in the form of small warts, as is the case of the *Amanita citrina*. The partial veil covers the cap and the tip of the stipe; as the cap expands, the veil breaks and its remains form a ring or annulus at the top of the stipe.

Amanita citrina, hemiangiocarpous.

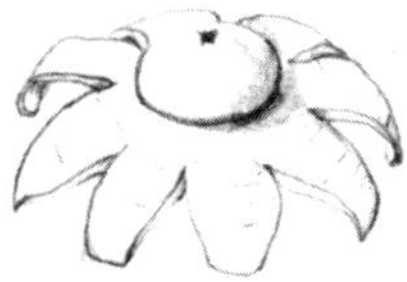

Geaster fimbriatus, angiocarpous.

Guepina helvelloides, gymnocarpous.

1. THE BASIDIOMYCETES

a. Gasteromycetes (Angiocarpous)

The Gasteromycetes are composed of a fertile gleba enclosed within the peridium, hence their name (g. *gaster:* belly). Although at the very beginning of their development they are all spherical or egg-shaped, they subsequently vary greatly in their morphology. In some cases the peridium is ruptured and a fragile and granulous stipe develops, which carries a rank and deliquescent gleba. This stipe could be either a simple cylinder, like that of *Mutinus*, or carry a cap at the top, as in *Phallus* or *Dictyophora*, which also displays an elegant frill. It can also look like a rather globular and coarse lattice, like that of *Clathrus*, or like an octopus spreading out its tentacles, as in *Anthurus*.

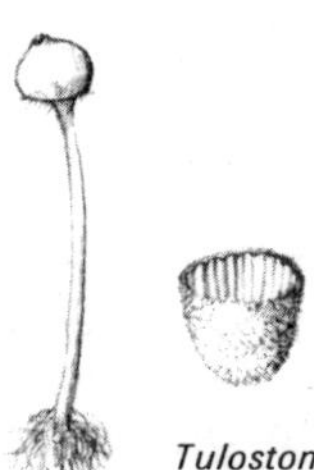

Tulostoma brumale, Cyathus striatus, angiocarpous.

Clathrus cancellatus, Dictyophora duplicata, Anthurus aseroiformis, angiocarpous.

The rest of the Gasteromycetes have a firmer flesh, which dries without rotting, and a shape which does not change with their development. Some of them are globular with a peridium which can be thin and fragile, as in *Bovista*, or thick and leathery as in *Scleroderma*. Others are pear-shaped, their spherical gleba being supported by a more or less conical sterile base: these are the genera *Lycoperdon*, if the top of the peridium shows, on maturity, a regular opening, and the *Calvatia* if the peridium breaks up irregularly. Sometimes this type of carpophore is supported by a long and slender stem, like the one of the genus *Tulostoma*.

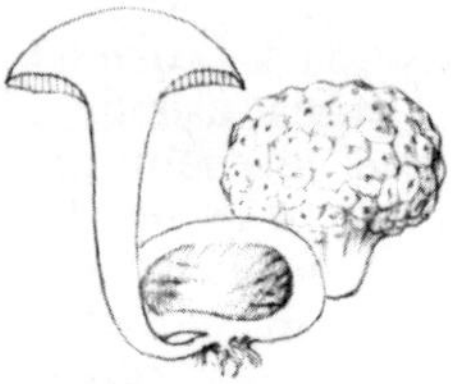

Scleroderma aurantium (with *Boletus parasiticus*) and *Lycoperdon coelatum*, angiocarpous.

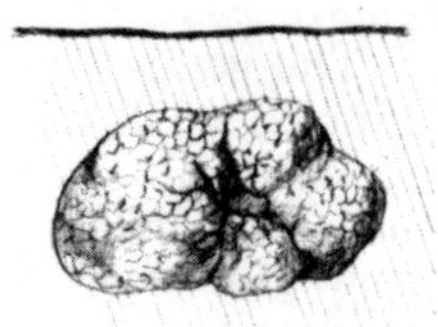

Tuber melanosporum, angiocarpous.

Even greater variations can be found in the carpophore which can, for example, look like a tiny bird nest complete with 'eggs' containing the spores: such is the case of the *Cyathus*. In others the peridium is composed of two membranes superimposed: the external one ruptures and rolls back to form a star supporting the internal, globular part; as in *Geastrum* and *Astraeus*. The star-shaped base of the latter is hygroscopic and folds back over the internal part in dry weather. Finally, some species of Gasteromycetes only develop under the surface of the soil: they are subterranean and resemble truffles, from which they are distinguished by their basidia: truffles are actually Ascomycetes.

b. Hymenomycetes (Gymnocarpous and Hemiangiocarpous)

Unlike the Gasteromycetes, gymnocarpous and hemiangiocarpous fungi carry their spore cell on the hymenium, hence their name of Hymenomycetes. The structures supporting the hymenium form the hymenophore (g. *phorus:* carrier) and their shape determines their classification into gilled, tubular, toothed (or spiked), wrinkled and smooth.

1. GILLED HYMENOPHORES (AGARICS)

The hymenium is situated on gills which radiate from the stipe towards the margins of the cap. The Agarics (l. *ager:* field) have these characteristics and are thus called because to this group belong the field mushrooms. They are divided into various genera and species classified according to the colour of their spores which can be established by a simple test: a well-developed and ripe cap is placed on a piece of white paper, gills downwards, and covered with a tumbler; it is then left for several hours (usually overnight) and when lifted the spore print should be clearly visible on the paper. If one expects the spores to be white or very pale one should use black paper. According to the colour of the spores, these mushrooms are therefore divided into melanosporous (g. *melas:* black), ianthinosporous (g. *ianthinos:* violet), ochrosporous (g. *okhra:* yellow, ochre), rhodosporous (g. *rhodon:* rose, pink) and leucosporous (g. *leukos:* white).

Melanosporous. The spores are black, or blackish/dark brownish, grey. The genus *Coprinus* contains some fifty species, usually of medium to small size. Their cycle is short and their gills soon turn into a liquid the colour of which is determined by that of spores. The genus *Gomphidius* also has blackish spores; the mushrooms belonging to this genus look like short and thick nails and, apart from the species *helveticus*, are covered by a mucilage; their gills are thick, widely spaced, of a waxy consistency and decurrent along the stipe.

Coprinus comatus.

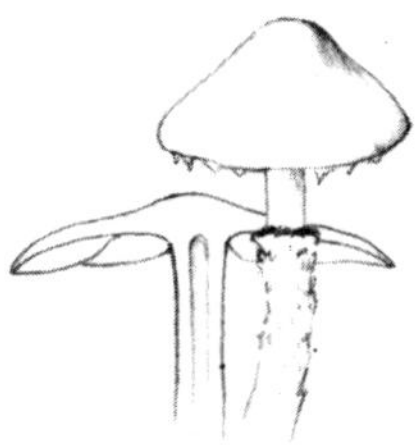

Stropharia aeruginosa.

Ianthinosporous. The genus *Stropharia*, containing some twelve species, has violet-brown spores. These mushrooms are often small and watery, their colours (yellow, green, maroon) fading quickly after cutting. Their stipe has a thin ring but no volva, and they grow either singly or in small groups of a few individuals.

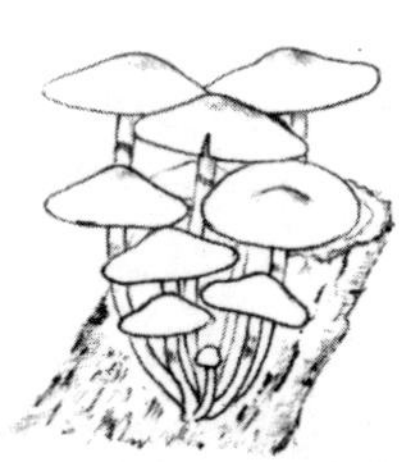

Nematoloma fasciculare.

Psalliota campestris.

The genus *Nematoloma* (syn. *Hypholoma*) also contains a dozen species whose spores are violet-grey. Again, they are quite small and they grow in groups on old wood. The genus *Psalliota* (syn. *Agaricus*) contains over sixty species with chocolate-coloured spores. Almost all the mushrooms of this genus are edible, like the famous Field mushroom (*Agaricus campester*) and Horse mushroom (*Agaricus arvensis*), as well as the cultivated mushroom common in the shops (*Agaricus bisporus*). They are terrestrial, with thin gills not reaching the stipe, and have a ring but no volva.

Ochrosporous. Five genera will be considered here. The genus *Paxillus* contains a few species of funnel-shaped mushrooms; the margin of the cap turns inwards towards the gills, which are thin, soft and decurrent along the stipe: they are also occasionally joined to one another, fragile and easily detachable from the cap, like the tubes of some boletes.

The genus *Pholiota* consists of some forty species of mushroom mostly growing on wood; they are caespitose, of varying dimensions, with non-decurrent gills; the cap is scaly, like the stipe which often bears a more or less visible ring. The *Hebeloma* are a genus of some sixty species. Their cap is viscid with involute margin. The gills are thickly set, thin, turning upwards towards the cap and light in colour. The stipe sometimes shows traces of a ring.

The *Inocybe* is a large genus of over 150 species, some toxic. They are terrestrial mushrooms, small, without ring or volva, their cap is conical, fibrillose or scaly, sometimes radially split; the stipe is fibrous and occasionally shows a pseudo-ring.

Pholiota destruens.

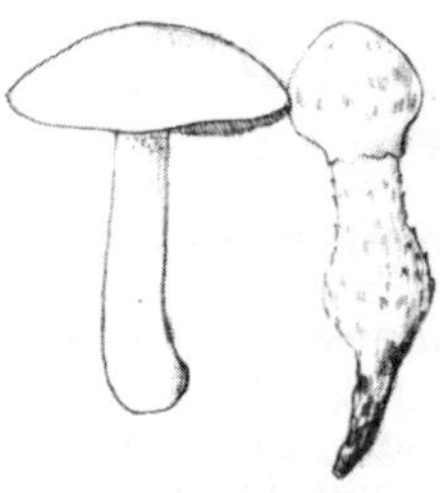

Hebeloma crustuliniforme (left) and *Hebeloma radicosum.*

Inocybe patouillardii.

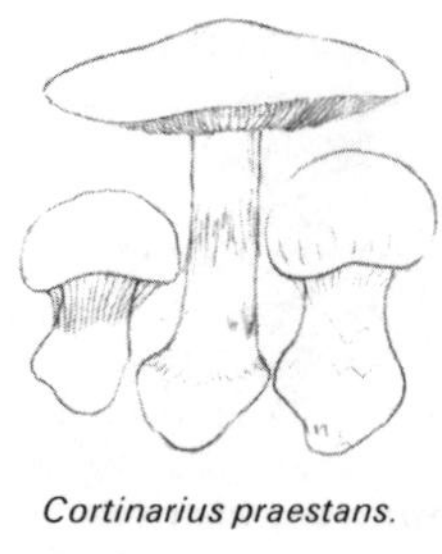

Cortinarius praestans.

A very large genus (almost 800 species) is *Cortinarius:* terrestrial mushrooms with a cobweb-like veil (cortina), which is sometimes mucilaginous, and non-decurrent gills. One, *Cortinarius orellanus*, is deadly; the other species have been variously grouped into subspecies, three of which are commonly accepted by mycologists: *Myxacium*, with a viscid cap and stipe; *Phlegmacium*, where only the cap is viscid; and *Cortinarius* strictly speaking, which is dry and lilac in colour.

Rhodosporous. The following genera have pinkish spores: *Volvaria*, a genus consisting of about 15 species; some are terrestrial, others grow on wood or other species of mushroom. The stipe ends with a bulbous, sometimes large, volva. The gills are wide, thin and free.

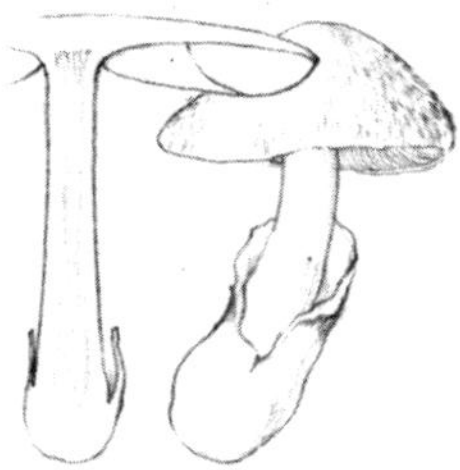

Volvaria bombycina.

Pluteus cervinus.

The genus *Pluteus* consists of about 50 species, generally growing on wood. The gills are free from the stipe and often completely detached from it. The stipe is separated from the cap.

The genus *Clitopilus* contains less than ten species of often conical mushrooms. Their cap is velvety with wavy or lobed margin, their stipe enlarged towards the cap and often eccentric, their gills decurrent along the stipe and pink in colour. The genus *Entoloma* is comprised of a few dozen species of smooth, naked mushrooms, with a cap which is thickly fleshed in the middle but getting thinner towards the edge and turning inwards towards the gills; the latter are not decurrent, but turn upwards towards the stipe and are hardly joined to it. Some species, eg *Entoloma lividum*, are toxic.

Leucosporous. Several hemiangiocarpous genera have white or pale spores. The main ones are here divided according to whether they are homogeneous (ie the tissue of the stipe is the same as that of the cap, and the two can only be separated by being broken apart) or heterogeneous (when the stipe tissue is different from the cap tissue and the two parts can easily be separated).

The following genera belong to the heterogeneous group. *Amanita* is the first, consisting of some fifty species. The gills are free from the stipe, which has both a ring and a volva. Amongst the species are the *Amanita caesarea*, which is excellent, and the deadly *Amanita phalloides, virosa* and *verna*.

About ten species form the genus *Amanitopsis*. They are similar to

Hygrophorus marzuolus (front), *Hygrophorus eburneus* (back, left), *Hygrocybe conica.*

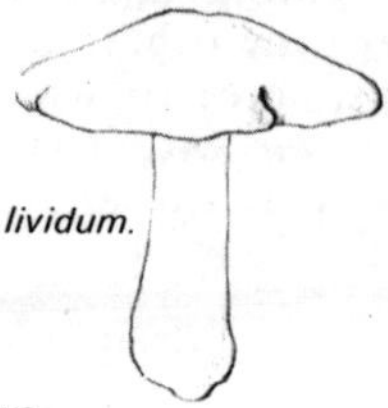

Entoloma lividum.

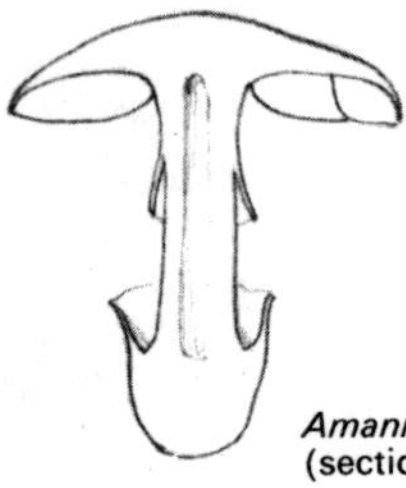

Amanita caesarea (section).

Amanita muscaria.

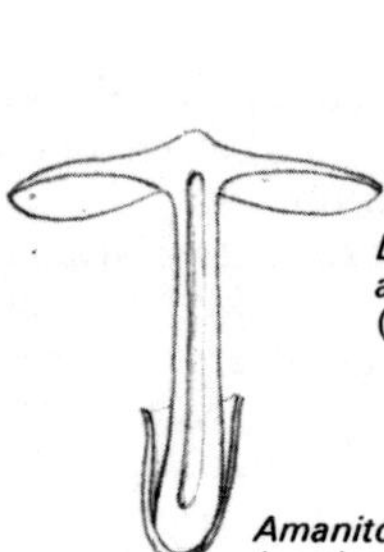

Amanitopsis vaginata (section).

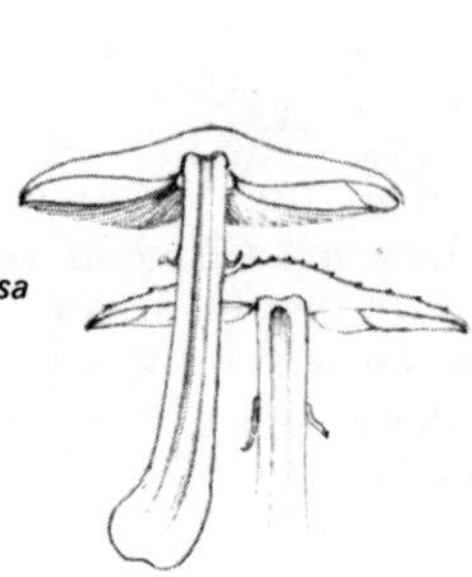

Lepiota excoriata (front) and *Lepiota acutesquamosa* (section).

Lepiota procera.

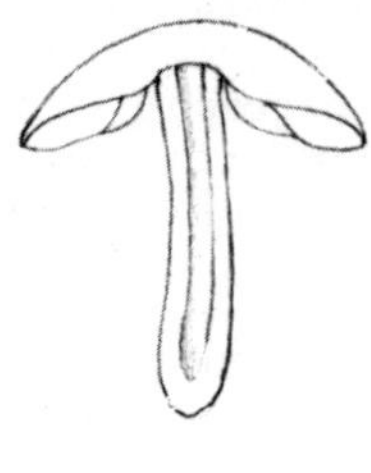

Lepiota naucina (section).

the amanites but do not have a ring; their stipe is slender and cylindrical, their cap not as wide as that of the amanites. They are often considered to be part of the genus *Amanita*.

The genus *Lepiota* is comprised of about forty species. These mushrooms have a ring but no volva and among them is the famous *Lepiota procera*, or Parasol mushroom, which can grow to a height of about 40cm.

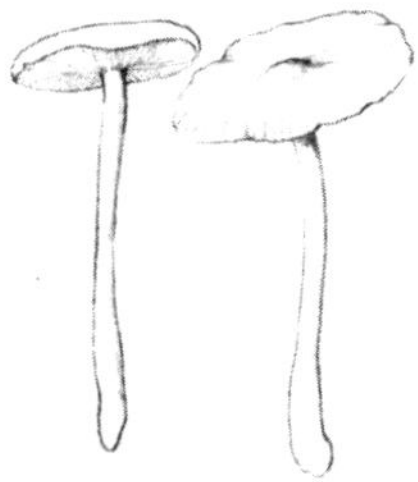

Collybia dryophila (left) and *Marasmius oreades*.

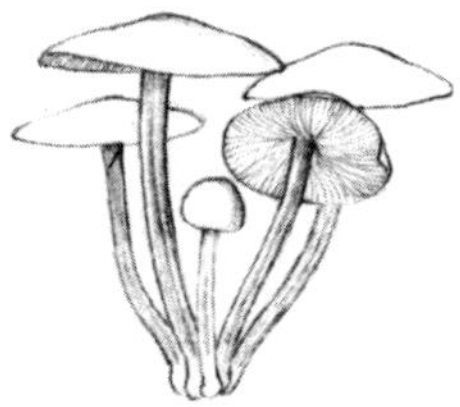

Collybia velutipes.

Collybia platyphylla.

Collybia is a genus consisting of about thirty species, some of which are small, and grow on wood in tufts, others are larger, terrestrial and can be found growing singly. They have neither ring nor volva, and their gills hardly touch the stipe at all. The winter-growing *Collybia velutipes* is edible and well worth trying. About thirty species form the genus *Marasmius*; some are terrestrial, some epiphyllic, some grow on wood remains. They have little flesh, no ring and no volva; many can be easily dried. Among them is the well-known *Marasmius oreades*, or Fairy-ring mushroom.

The homogeneous leucosporous mushrooms can be divided into three groups based on whether their flesh is granular or not. Mushrooms with a granular, and therefore fragile, flesh belong to the genus *Lactarius* (about 100 species) and *Russula* (over 100 species); the former exude milk when broken, unlike the latter. The mushrooms with non-granular flesh are subdivided into terrestrial and wood-growing. The latter include not only the mushrooms growing on plants whether alive or dead, but also those growing on dead branches, loppings and prunings. This group includes the genus *Armillaria*, with a ring normally upturned and only later

Lactarius volemus.

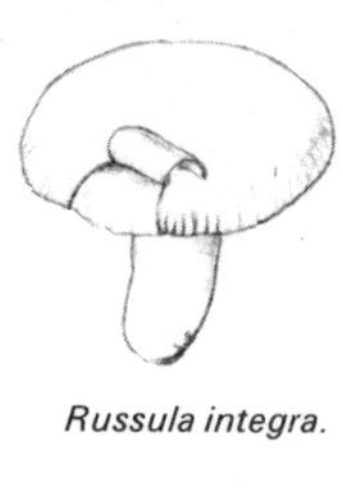

Russula integra.

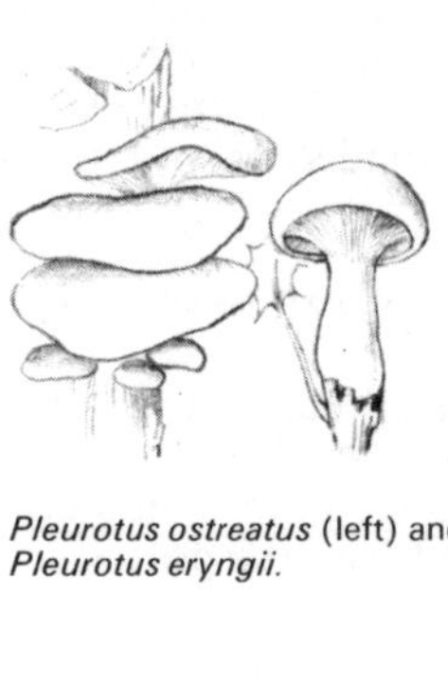

Pleurotus ostreatus (left) and *Pleurotus eryngii*.

rather more floppy, and gills slightly decurrent. Some mycologists ascribe to this genus only the *Armillaria mellea* (Honey fungus). *Pleurotus* is another genus comprised of about ten species. They have little or no ring, decurrent gills, regular margin and a stipe which can be eccentric, lateral or totally absent.

The genus *Lentinus* comprises some ten species with rather firm flesh; they are distinguished from the previous genus by the dented edge of the gills and can be dried without rotting. *Panus* is a similar genus, consisting of a few species, and with regular edge on the gills.

The terrestrial genera are five. *Hygrophorus* contains some 100 species: their gills are waxy, thick and juicy, and easily detachable from the cap. Three sub-genera are normally ascribed to this genus: *Limacius*, with decurrent gills and fleshy, viscid cap; *Camarophyllus*, also with decurrent gills, and cap which is fleshy but dry; *Hygrocybe*, with non-decurrent gills and often thin cap, and usually a fruit-body of rather small dimensions.

The genus *Clytocybe* also contains some 100 species; their gills are decurrent, not waxy, thin and crowded.

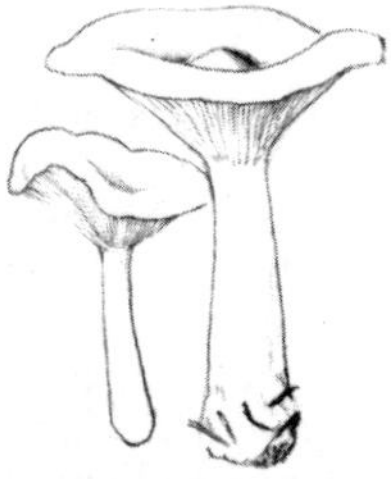

Clitocybe dealbata (left) and *Clytocybe geotropa.*

Tricholoma rutilans.

Again about 100 species are part of the genus *Tricholoma*; the gills are either adnexed or sinuate and the fruit-bodies are thick and fleshy, often quite bulky. Worthy of a mention are the genera *Mycena* and *Laccaria*, the former consisting of over 150 species, the latter of very few. These mushrooms are often small, of pleasant colour, with neither ring nor volva. The gills of the *Laccaria* are distant, deep and slightly decurrent, those of the *Mycena* crowded and non-decurrent. Some species of this latter genus grow on wood, some exude milk, some are heterogeneous rather than homogeneous. It is obviously a very complex genus which should really be subdivided or split into several more.

2. TUBULAR HYMENOPHORE (BOLETES AND POLYPORES)

The hymenophore of these species consists of innumerable tiny tubes, lined on the inside by the hymenium; the spores are let out through a pore on their tips. These characteristics are typical of the boleti and the polypores.

The boleti are fleshy mushrooms, rather soft, with stipe and cap, fairly similar to the gilled mushrooms we have seen above. With the exception of *Boletus parasiticus*, which grows on the genus Scleroderma, they are terrestrial; the tubes grow on the underside of the cap forming a squashy conglomerate which can be separated all in one piece. There are more than 100 species distributed over several genera and sub-genera: *Suillus* (syn. *Ixocomus*) with a viscous cap; *Xerocomus*, with a dry and velvet cap; *Boletus* strictly speaking, consisting of a number of large species and having stipes which are either dotted or covered in coloured web-like tissue; *Leccinum*, with their rather tough flesh and scaly stipe; *Gyrodon*, whose tubes are very short and difficult to separate; *Gyroporus*, with whitish spores and hollow stipe; *Tylopilus* with pink spores and very bitter flesh; *Porphyrellus*, chocolate-brown all over; and finally the odd *Strobilomyces* with its cap covered in darkish scales which make it look like a pine cone; it tends to dry rather than rot.

Boletus elegans (front) and *Boletus edulis.*

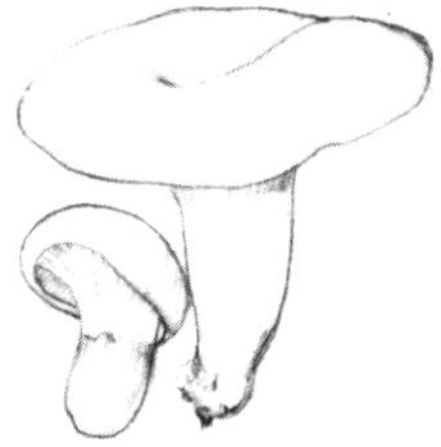

Paxillus atrotomentosus (left, immature) and *Paxillus involutus.*

Mycologists often associate three other genera of gilled mushrooms with boletes: *Phylloporus*, with thick gills, distant, irregular and occasionally joining together to form variously-sized hives (hence their name: g. *phullon:* leaf, gill, and *poros:* pore); and *Gomphidius* and *Paxillus* already dealt with.

Although thoroughly different from the boletes, the polypores also have a tubular hymenophore. They grow mainly on wood and seldom have a stipe. Their flesh is often as leathery as cork or as tough as wood. There are several hundreds of species divided among numerous genera, but we can only look at a few of them. The genera *Ungulina* and *Ganoderma* have large and woody fruit-bodies, with a crust on the upper surface. The former have white spores, the latter rust-coloured ones. The genus *Phellinus* has the same kind of fruit-body and white spores but no crust. The species of the genus *Trametes* are slightly smaller with white and porous flesh. The *Phaeolus* are more tender, with rusty flesh which becomes a beautiful purple when exposed to ammonia vapour. Only one species belongs to the genus *Fistulina–F. hepatica*, characterised by a flesh as fragile as that of gilled mushrooms and by its tubes, juxtaposed without being welded. It is commonly known as

Phellinus robustus.

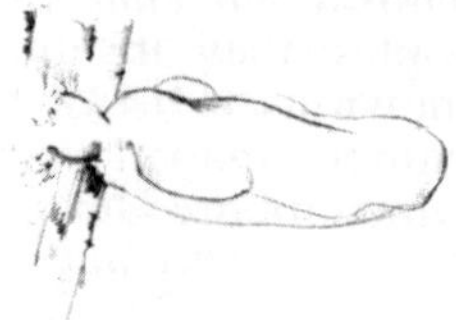

Fistulina hepatica.

Beefsteak fungus and is the only polypore which can be eaten. The genus *Polyporus* also consists of very fragile species; these mushrooms are sometimes supported by a stipe which can be forked. One of these species, *Meripilus giganteus*, grows on stumps and has carpophores arranged like the petals of a flower (or overlapping, fan-shaped brackets), the diameter of which can be up to a metre. The species of the genus *Melanopus* have stipes turning black at the base (hence their name: g. *melas:* black, *pous:* foot); the flesh of the carpophore is rather stringy.

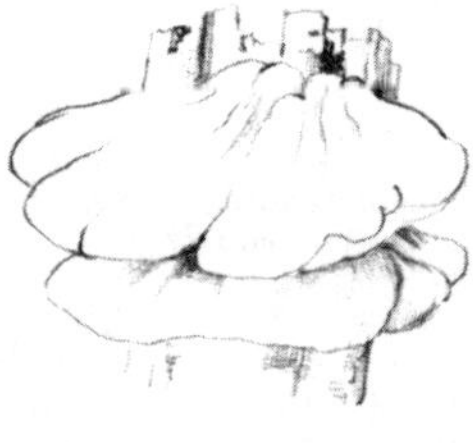

Polyporus sulphureus.

Sarcodon repardum (front), *Tremellodon gelatinosum* (back).

3. TOOTHED HYMENOPHORE (HYDNUMS)

Some species of mushroom, although not many, carry the hymenium on the surface of spikes or teeth growing downwards. The genus *Hydnum* has mushrooms with soft, white and edible flesh, particularly *Hydnum repandum*, or Wood-hedgehog. Other genera, such as *Sarcodon*, have a coloured flesh, fairly thick and often bitter, while the *Calodon* is more leathery. There is also the wood-growing *Dryodon* and the *Auriscalpium vulgare*, or Ear-pick fungus, with its kidney-shaped cap on a slim stipe growing on pine cones.

4. WRINKLED HYMENOPHORE (CHANTERELLES)

In these species the hymenium is found on the surface of more or less deep wrinkles. The genus *Cantharellus*, consisting of five species, has wrinkles similar to gills with ramifications; *Cantharellus cibarius*, the well-known chanterelle, is very good to eat. In the genus *Craterellus* the folds are much more shallow, often like ripples; they are totally absent in the *Craterellus cornucopioides*, known as Horn of Plenty or Trumpet of the Dead owing to its trumpet-shaped fruit-body. In the genus *Nevrophyllum* the hymenophore is almost smooth and the carpophores purplish.

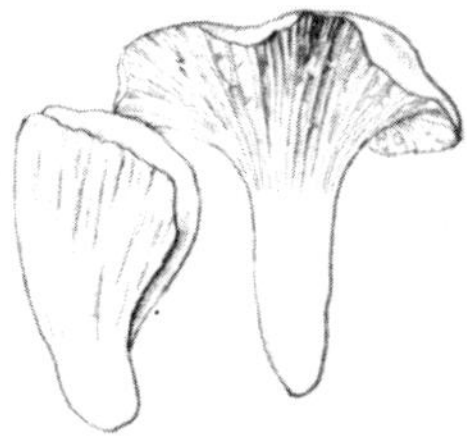

Cantharellus cibarius (right) and *Nevrophyllum clavatum.*

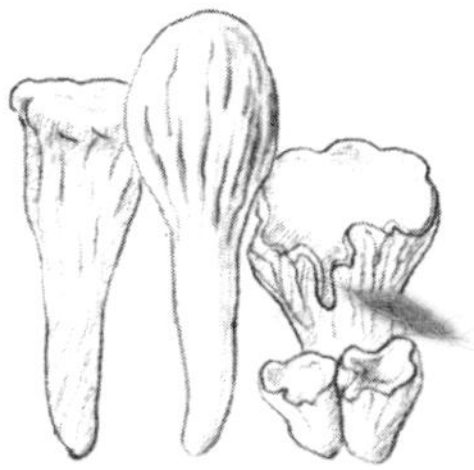

Clavaria truncata, Clavaria pistillaris and *Nevrophyllum clavatum.*

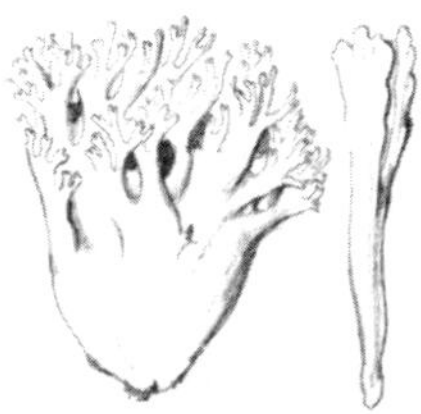

Clavaria formosa.

5. SMOOTH HYMENOPHORE (CLAVARIAS AND CORTICIUMS)

This is the simplest type of hymenophore: the hymenium covers totally or in part the surface of the carpophore. The Basidiomycetes thus characterised are divided into two large groups: the first are the *Clavaria*. Their fruit-bodies are often shaped like clubs, awls, corals, or tiny, leafless shrubs with a varying number of ramifications. Their flesh is tender like that of gilled mushrooms and their spores are white or beige. One species, *Sparassis crispa*, is a globe of flat and wavy blades which earned it its name of Cauliflower fungus; it grows under conifers and is edible. No *Clavaria* is dangerous but some species, among them the *Ramaria formosa* with its large pink branching carpophore, are highly purgative.

The second group of species with a smooth hymenium consists of wood-growing fungi which have the appearance of crusts on the surface of branches and trunks. The various species are difficult to identify, the genus *Stereum* being the best known. In some species, like *Stereum rugosum*, the hymenium becomes a deep red when stroked.

c. Heterobasidiae (Gymnocarps)

Our survey of the Basidiomycetes should not leave out three small groups which the mycologists have separated from those already mentioned owing to their basidia (partitioned or forked). Their carpophores are gelatinous or rubbery, fragile or hard, and they mostly grow on wood. They are the Tremellales, Auriculariales and Dacrymycetales.

The Tremellales consist of a number of species with a gelatinous carpophore, sometimes soft, sometimes elastic. The species of the genus *Tremella* grow on wood and are rather fragile, their bodies presenting irregular lobes, brain-like and yellow or light brown. The fungi of the genus *Exidia* have a harder body, also grow on wood, and are normally dark brown or black. *Guepinia helvelloides*, on the other hand, is terrestrial, rather rubbery, shaped like a tongue folded into a spout, and usually red or rusty. Finally, *Pseudohydnum gelatinosum* is greyish and entirely gelatinous; the underside of its cap is punctuated by teeth, and it is found growing on the stumps of conifers.

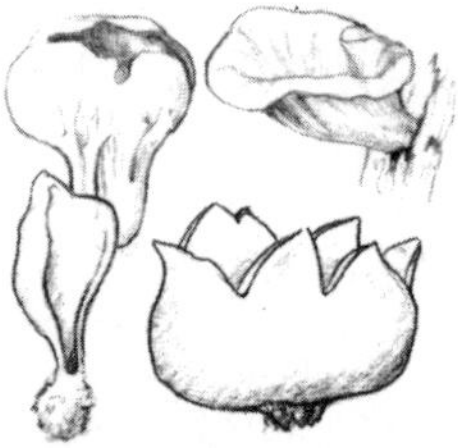

Acetabula vulgaris and *Hirneola auricula-judae* (above), *Otidea onotica* and *Sarcosphaera eximia* (below)

The one species among the many of the Auriculariales which we will look at is *Hirneola auricula-judae*, the Jew's ear fungus, so called because of its lobed shape looking just like an ear. It grows on wood, is rather rubbery and dark brown. The most common species of Dacrymycetales is the *Calocera viscosa*, shaped like a club and yellow; its skin is viscous but the flesh rather tough. It grows frequently on the stumps of conifers.

2. THE ASCOMYCETES

a. Pezizales

All the species of this group are gymnocarpes; their asci are contained by an hymenium which covers part of the carpophore. The carpophores of the genus *Morchella* (Morels) are composed of a whitish stipe and a fertile cap, creased and irregularly honeycombed, looking a bit like a sponge. They grow in spring, are excellent to eat, and are often sought after. They do contain toxins, but these are fortunately destroyed during the process of cooking.

The genus *Gyromitra* contains only three species, the best known of which is *G. esculenta*, with its whitish stipe and brain-shaped fertile cap. It grows in March and April and is also good to eat, although less savoury than the morels. It also contains toxins which are destroyed in the cooking but are nonetheless more dangerous than those of the morels: one should always cook them carefully and throw away the water or juices.

The genus *Helvella* (False morels) has carpophores often composed of a stipe, deeply ribbed or furrowed, and a fertile cap, cup-shaped or saddle-shaped and irregularly folded. They are edible, when cooked, but extremely dull.

One of the important groups within this family is that of the Pezizales. Their cap is like a cup of varying dimensions: according to the species, they can be 1–20cm in diameter; the hymenium adheres to the upper surface. The most common species is the *Aleuria aurantia* (syn. *Peziza aurantia*), the Orange-peel peziza, which is edible when raw. *Sarcosphaera eximia* has a lilac hymenium and contains toxins which are destroyed by cooking.

Helvella monachella (left) and *Morchella rotunda*.

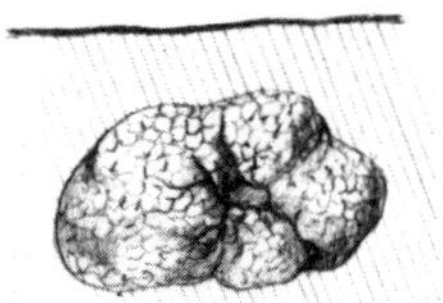

Tuber melanosporum.

b. Tuberales (Truffles)

Truffles are globular, subterranean fungi, whose spores form within a gleba inside the carpophore; they are angiocarps. *Tuber melanosporum*, the commercial truffle used in France, is the most highly rated from the point of view of edibility. Its taste is very strong and aromatic and it is often used as a condiment. Many other truffles are edible: *T. brumale*, which grows in winter, for instance, although it is a lot less appreciated. There are in all about 100 species of truffles, spread over several genera, but most of them are not worth eating at all.

GENERAL CLASSIFICATION TABLE

BASIODIOMYCETES

Gasteromycetes (angiocarps)
Mutinus, Phallus, Dictyophora, Clathrus, Anthurus, Bovista, Scleroderma, Lycoperdon, Calvatia, Tulostoma, Cyathus, Geastrum, Astraeus

Hymenomycetes (gymo- and hemiangiocarps)

GILLED HYMENOPHORE (AGARICS)
Melanosporous: *Coprinus, Panaeolus,* (*Gomphidius*)
Ianthinosporous: *Stropharia, Hypholoma, Agaricus*
Ochrosporous: *Pholiota, Hebeloma, Inocybe, Cortinarius,* (*Paxillus*)
Rhodosporous: *Volvaria, Clitopilus, Entoloma, Pluteus*
Leucosporous: *Amanita, Lepiota, Armillariella, Lactarius, Russula, Pleurotus, Panus, Lentinus, Hygrophorus, Tricholoma, Clitocybe, Marasmius, Collybia, Laccaria, Mycena*

TUBULAR HYMENOPHORE (BOLETI AND POLYPORES)
Boleti: *Suillus, Xerocomus, Boletus, Gyrodon, Gyroporus, Tylopilus, Porphyrellus, Gomphidius, Paxillus*
Polypores: *Ungulina, Ganoderma, Phellinus, Trametes, Phaeolus, Fistulina, Polyporus, Melanopus*

TOOTHED HYMENOPHORE (HYDNUMS)
Hydnum, Sarcodon, Dryodon, Auriscalpium

WRINKLED HYMENOPHORE (CLAVARIAS AND CORTICIUMS)
Clavarias: *Clavaria, Ramaria, Sparassis*
Corticiums: *Stereum*

Heterobasidiae
Tremellales: *Tremella, Exidia, Guepina, Pseudohydnum*
Auriculariales: *Hirneola*
Dacrymycetales: *Calocera*

ASCOMYCETES

Pezizales (gymnocarps)
Morchella, Gyromitra, Helvella, Peziza, Sarcosphaera

Tuberales (angiocarps)
Tuber

B. THE LIFE-SPAN OF MUSHROOMS

Mushrooms are plants without chlorophyll—the substance that makes leaves green and by which plants transform carbon dioxide and water into the substances they require to survive. Mushrooms must extract this carbon dioxide from other organisms, whether alive or dead. In short, they feed on organic carbon dioxide. Some of them attack live plants, feed on them, kill them and go on feeding on them: they are parasites (g. *parasitos:* feeding with) of the live plant and saprofites (g. *sapros:* putrefying, and *fiton:* plant) of the dead one. *Armillaria mellea* and *Polyporus sulphureus* are two such examples.

Other fungi form an association between their mycelium and the plant roots called a mycorrhyza (g. *mukes:* mushroom, and *rhiza:* root). Through this symbiosis the fungus supplies the plant with useful substances while at the same time deriving those it needs. Some plants could not live at all without the co-operation of these fungi, and it is therefore necessary to respect them all, even the poisonous ones.

As we know, chlorophyll needs sunlight to initiate photosynthesis; it follows that green plants could not live without sunlight. Fungi, on the other hand, do not need it quite so much; only their colouring is affected, being more vivid and intense where there is more light, and paler in shadier places. What is very important to their survival is temperature and humidity; not so much the air temperature as that of the soil in which the mycelium spreads. We all know that the soil absorbs and retains temperature in direct proportion to the darkness of its colour and its moisture retention: such a soil is therefore much more favourable to the development of fungi than a light, sandy and dry one.

This said, there are fungi which grow and ripen under the snow and even in the hot sands of deserts. Nevertheless, the majority of them do need a temperate and fairly humid climate, such as is found in most of Europe in late summer and autumn. Summer storms and the first autumn rains usually forecast a good crop; however, some fungi only fruit some time after the rain, when the subsoil begins to dry up: then, close to inactivity or death, they decide to perpetuate the species and produce their carpophores full of spores. That is why mushrooms are said to fear the wind! Persistent wind quickly dehydrates the soil and impedes the growth of the mycelium, causing it to be dormant or greatly reduce the production of carpophores. And do fungi really grow so quickly that you can watch them do so? Some indeed do have a very brief, and therefore quick, cycle: they are those fungi, watery and thin, which generally grow on manure (fimicolous, l. *fimus:* dung, *colo:* inhabit), whose carpophores raise their little caps on slender stipes at sunrise and by sunset they wilt, already almost unrecognisable. This happens to mushrooms belonging to the genus *Coprinus*, among them *Coprinus comatus* which is excellent to eat. Sometimes the carpophores appear suddenly. The author has witnessed the eruption (no better term) of a tuft of white, huge carpophores belonging to *Amanita*

Coprinus comatus.

Amanita ovoidea.

ovoidea: there were four or five of them, the smaller ones surrounding a central cap about 20cm wide. And eruption it certainly was, because all of a sudden the calcareous and well-trodden soil of the path cracked and broke up like the crater of a volcano and out came those splendid mushrooms! Obviously they had long been produced by the mycelium but only then had they grown strong enough to break the barriers that restricted them, emerge in a compact group and quickly open, expand and grow.

Some fungi can extend their vegetative period for weeks and months and rest only during cold spells; then, when the more clement weather comes, they resume their development, and this continues for years. They are usually wood-growing fungi, the older part of their carpophore being woody or cork-like. Other fungi die every year, at the end of their short growing season, and have to grow again at the beginning of the next season from the spores spread by the previous generation. In other species the mycelium renews itself year after year, spreading underground in concentric circles; out of these circles the carpophores are born each new season, giving rise to the folk tales of fairy rings or witches' circles. Within these rings the dead mycelium fertilises the grass, which grows taller and greener.

C. THE HABITAT

The majority of fungi, particularly those used for culinary purposes, grow in woods; a smaller number in the meadows; even less in waste grounds, rocky and sterile soils; and very few where chemical fertilisers have been used. As we have seen above, some species grow on the sands of deserts, on salty beaches, on old stable or barn beams, on the ceiling of cellars; some under dead leaves, in a wood, or even under snow. The majority of fruit-bodies emerge from the substratum where the mycelium lives and are clearly visible in the open air; a few species produce carpophores within the substratum and therefore usually underground.

Oak and chestnut woods are beloved of the *Amanita caesarea*, Caesar's mushroom; mixed woodlands containing conifers, whether on mountains or plains, are the home of the deadly *Amanita phalloides*, usually greenish. The much sought-after *Boletus aereus* is usually found in broadleaved woods, while numerous tufts of *Boletinus cavipes*, *Boletus elegans* and *B. viscidus* grow at the foot of larches, in a sunny position. *Boletus bovinus*, *B. luteus* and *B. variegatus* grow under pine trees; *Boletus granulatus* likes the foot of young pine trees and there it can mature several times in the same season, while *B. luteus* needs rather old pine trees. *Boletus carpini* grows under hornbeams, *B. leucophaeus* and *B. rufescens* under birches, *B. duriusculus* under poplars, *B. crocipodius* under oaks and beeches, and *B. aurantiacus* under hornbeams, birches and poplars but rarely under oaks and beeches.

All these species grow on mycorrhiza and are therefore useful to many trees of our woodlands. It is the parasitic species, like *Polyporus sulphureus* and *Armillaria mellea*, which do the damage. Other fungi are parasitic without causing damage: these are the species that live off other fungi, like *Boletus parasiticus* which grows on *Scleroderma aurantium*; *Nyctalis parasitica* which grows on *Russula nigricans* and on various species of the genus *Lactarius*; *Nyctalis asterophora* which grows on *Lactarius vellereus*; and *Volvaria loveiana* which grows on *Clitocybe nebularis* as well as on *C. clavipes* and on various species of the genus *Tricholoma*.

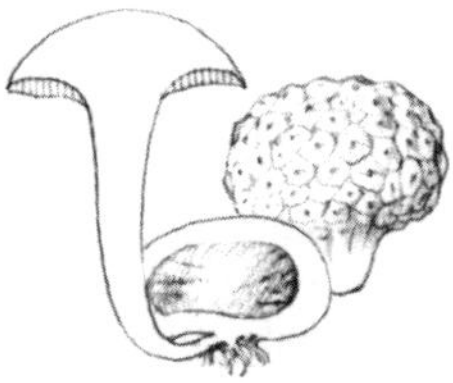

Scleroderma aurantium with *Boletus parasiticus* (front) and *Lycoperdon coelatum*.

Nyctalis asterophora on *Russula nigricans*.

In the days when charcoal was still produced by charcoal-burners in the woods, it was easy to find the edible *Cantharellus carbonarius* in the spots where the wood for burning had been piled up. Now it is sometimes found after woodland fires. Along woodland paths, even on trodden earth, one can find some species of *Peziza*, among them, although rarely, the large, purplish *Peziza* (or *Sarcosphaera*) *eximia*. *Marasmus oreades* grows along field paths, in meadows and pastures, where one can also find various species of *Psalliota*, while other species of the same genus grow within the woods. The genera *Lactarius* and *Russula* grow almost exclusively in the woods, while many species of the genus *Hygrophorus* grow both in woods and meadows. The genus *Lepiota* contains some species which are thoroughly ubiquitous: they can grow on any kind of soil, whether cultivated or not, deep and humid or stony and dry.

Loose, sandy soils, rich in calcareous and organic substances, are home to the tufts of the excellent *Coprinus comatus*, while *Coprinus atramentarius*, edible and tasty provided certain precautions are taken, prefers heavily manured soils. *Volvaria gloiocephala* and *Stropharia ferrii* grow on heaps of decomposing organic matter, like straw and sticks mixed with soil; *Strobilomyces strobilaceus* grows on soil disturbed by moles, in a sunny spot; while species of the genus *Morchella* are found in spring on soil produced by excavations, particularly if it is sandy, moist and open to the sun. *Psalliota bernardi*, *Bovista plumbea* and *Geaster nanus* manage to grow on salty and sandy beaches by the sea.

Hygrophorus puniceus (section).

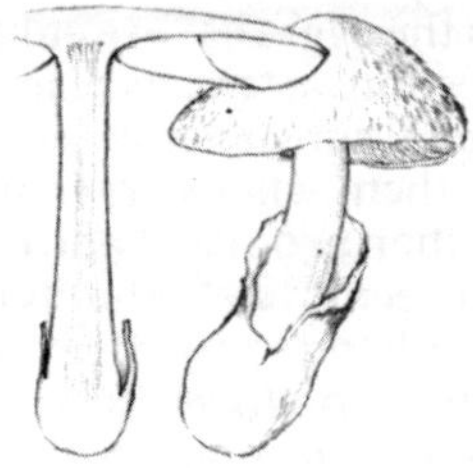

Volvaria bombycina.

Some fungi can, up to a point, vary their characteristics to adapt to habitat variations. *Cantharellus cibarius*, for instance, when in association with oaks, grows in early spring and is orange-yellow and of medium size; under the beeches it ripens later, and is larger but of a paler colour; under the hornbeams it grows very late in tufts and is very thin; under fir trees it is thick, pale and slightly bitter. *Armillaria mellea* is another such fungus, changing its character according to the tree it lives on: honey-coloured on mulberries, cinnamon-coloured on poplars, ochre on oaks, olive-brown on acacias, reddish-brown on conifers.

Since fungi have well-defined needs as to temperature, humidity and substratum, it is understandable that they could not grow beyond a certain altitude and latitude. Species like *Polyporus ovinus*, quite happy in the Alps, will only be found on the plains at a more northerly latitude; by the same token, only a few hundred species can grow in the Alps above 2000m. Among the commonest species in the world is the excellent *Boletus edulis*, which grows both on the mountainside and on plains, west or east. Its immediate rival in edibility, *Amanita caesarea*, is much more difficult to find as one travels north; in fact it becomes more difficult to find as *Amanita muscaria*, also red but deadly, becomes easier.

D. THE CULTIVATION OF FUNGI

One stands a better chance of finding any given fungus if its habitat is known, and even then there will be competition from other collectors or gourmets. Hence the need, since ancient times, to prepare an artificial habitat suitable to the cultivation of certain edible species. Animals thought about this first; not anthropomorphic animals but insects: the ants. The leafcutter ants of Brazil grow a species close to the genus *Amanita* and feed on the plasma which fills parts of its mycelium. Obviously this particular species of fungus can only produce a fruit-body on abandoned ant-hills. Other ants of equatorial America, on the other hand, live on the carpophores and submit themselves to a lot of work in order to grow enough of them. First they choose the right soil, then they move into it parts of mycelium removed from the previous, and now exhausted, area and place them in the right spots. Only when all this transplanting is done does the queen begin to lay eggs; if necessary, it is the eggs that are eaten rather than touch the precious mycelium. As soon as the larvae have become workers, they look for the vegetal substances needed by the mushrooms, shred them finely and put them where the hyphae can best utilise them; the carpophores are then produced and the ants reap their crop.

It seems likely that the cultivation of mushrooms by men, on the other hand, only happened by chance—rather as it happened to the man who decided to rid his plot of mushrooms of the genus *Gomphidius* by throwing them on the compost heap: the following year, the plot, duly fertilised with compost, produced unbelievable quantities of those very mushrooms. What is not clear is when men began to cultivate mushrooms. The Bible gives no evident reference to mushrooms at all, which is surprising as in those lands, during and immediately after the rainy season, mushrooms are abundant and were surely used by people, particularly during lean periods. Among the mushrooms common in the Middle East many are edible and tasty: *Tricholoma terreum, Boletus granulatus* and *Amanita ovoidea.*

The Greeks and the Romans grew *Agrocybe aegirita* by spreading the powder of the poplar's bark on rich humus in the vicinity of poplars: the wind would then disseminate the spores on the prepared ground. Nowadays, one saws poplar trunks into rings, places them on the ground, spreads *Agrocybe* gills on to them, covers them with a thin layer of soil and keeps them moist. Some even place a paste of *Agrocybe* caps underneath the bark and wait. The first carpophores should appear within nine or ten months, should grow several times during the same season, and come up again for several seasons until the lignin is completely consumed by the mycelium. Thus *Volvaria bombycina, Pholiota mutabilis, Collybia velupites, Armillaria mellea* and a few other species are cultivated. Among the terrestrial species also in cultivation in Europe are *Coprinus comatus, Clitopilus prunulus, Rhodopaxillus nudus* and *Craterellus cornucopioides.* Thanks to advanced

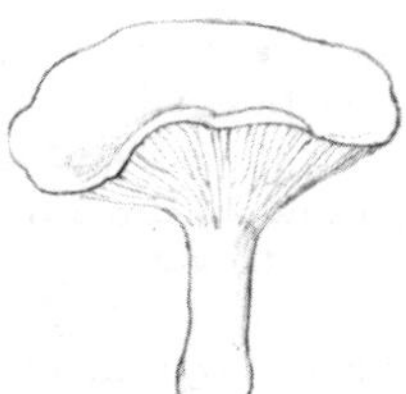

Clitopilus prunulus.

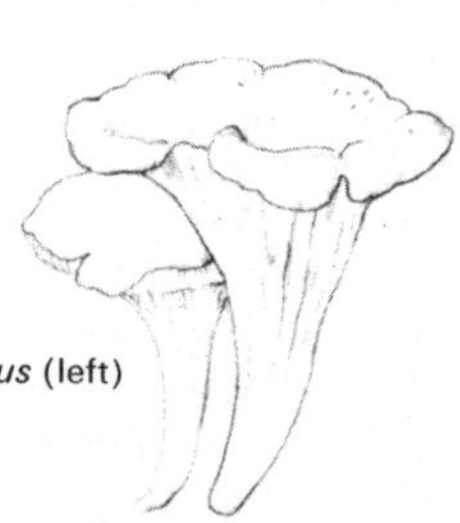

Cantharellus cibarius (left) and *Craterellus cornucopioides.*

knowledge of the symbiotic process, many species are now cultivated on a vast scale: they all belong to the genera *Amanita, Tricholoma, Cortinarius, Lactarius, Gomphidius, Paxillus, Boletus* and *Tuber*. In India *Coprinus niveus* is grown on dung; *Lentinus* (or *Tricholomopsis*) *edodes* is grown on the oak *Pasania* by the Japanese, who call it 'Shiitake', ie oak fungus; and a couple of species of *Psalliota,* usually *Psalliota bispora*, are commonly grown everywhere as the well-known champignons.

Psalliota bispora could be grown in the open air, but it is preferable to use confined environments which allow a careful control of humidity, temperature and ventilation. Horse manure, rich in lignin, is normally used: it should be well rotted and almost odourless, and of a rich and homogeneous consistency. It is then spread on shelves like thick mattresses, shallow holes are opened on the surface and within them a piece of mycelium is placed (this can be bought from specialist growers). Within three weeks the hyphae will appear and should be covered with a thin layer of sand, about 3cm thick. The first carpophores will begin to show a few weeks later and will continue to grow for a few months, until the mycelium has exhausted the lignin of the manure; this should then be replaced. The room temperature at the beginning should be around 19°C and the relative air humidity 85%. After the first month of cropping, temperature should be lowered gradually to 13°C and the humidity raised to 90%. Ventilation should be provided all the time. When grown commercially, all this is of course automatic.

E. PICKING AND COLLECTING MUSHROOMS

Cultivated mushrooms are only picked when of medium size; when small they are uneconomical to resell and when fully grown less appealing and tasty. This should also apply to mushroom-picking in the wild: neither too early nor too late. The young ones should be left to grow and the old ones to look after the reproduction of the species. One should also start early in the day, although not while the woods are still too dark; true, some species are luminescent, *Clitocybe olearia* for instance, but it is poisonous, and at any rate it is just about strong enough to light itself and not the rest of the wood! One should also avoid mushrooms which have been soaked by rain or have been frozen by the first autumn frosts; or those which are clearly infested by larvae or insects.

When picking mushrooms one should avoid damaging the mycelium; one should really cut them at the foot, particularly in the case of *Sarcodon repandum*, very much sought after, in order to let other fruit-bodies grow around the foot of the old one within a few weeks. This is also the easiest way to cut some wood-growing fungi, such as *Pholiota mutabilis, Collybia velutipes, Armillaria mellea*, and others. Rather than place the cut mushrooms in plastic bags, always lay them, cleaned of any particle of soil or sand, into baskets, to avoid their getting bruised or otherwise damaged. Should a mushroom be suspected of being host to larvae, it should be placed in the basket cap downwards: in a few hours all the larvae will have moved upwards into the stipe, which can then be easily removed.

Do not use water to clean the mushrooms as you will spoil their flavour. Brush them lightly with a sponge or an old toothbrush; use a sharp knife to remove the bitter skin of the *Polyporus confluens* and others which are not easily peeled; you can peel *Boletus viscidus, Boletus luteus, Boletus granulatus, Gomphidius viscidus* and *Gomphidius glutinosus* with your fingers. You would have to remove the scales from the cap of *Lepiota procera* taking care not to damage the cap itself. As for other mushrooms, like the best of the boleti, the skin should be left on as it provides a great part of their flavour.

While the gourmet's basket should contain only those mushrooms he feels one hundred percent sure about, that of the mycologist will show a collection of poisonous or inedible fungi and any as yet unknown to him. It is obviously easier to identify and classify a fungus if several specimens of the same variety are available at various stages of development. Fungi picked for this purpose should not be cleaned: they should still show part of the mycelium and should be placed in individual plastic bags accurately numbered, and then in the basket. The serious collector should then write in a notebook, under the corresponding number, all information necessary to the identification or further knowledge of the specimen he picked: for instance, when and where it was found, the weather, the surrounding vegetation, the altitude, the colour and smell of the fungus, and even its taste, taking care not to swallow any of it, and remembering that some species quickly alter their colour, taste and smell once picked.

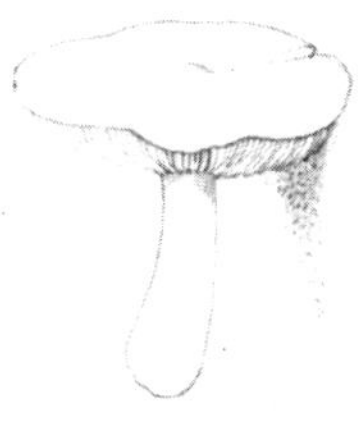

Gilled hymenium in *Russula caerulea.*

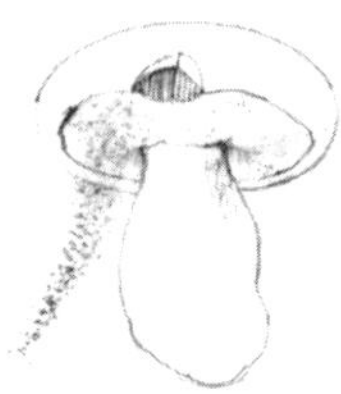

Tubular hymenium in *Boletus aereus.*

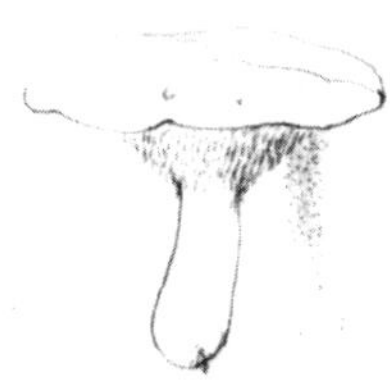

Toothed hymenium in *Sarcodon repandum.*

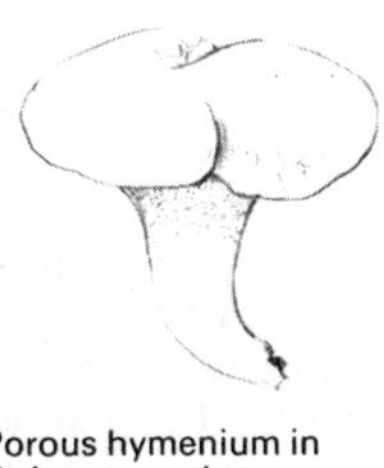

Porous hymenium in *Polyporus ovinus.*

Clitopilus prunulus

Smooth hymenium in *Guepina helvelloides.*

The colour of the spores, as we have seen, is very important for the identification of fungi; so are their shape, decoration, content and dimensions. We have seen how to obtain a spore print and the microscopic characteristic will be dealt with in due course. Let us look at another important factor in the identification process, ie the fungus' morphology. Always take into consideration the fungus' shape, its colour, its ornaments, the structure of the flesh of both cap and stipe: whether the flesh contains milk, whether it discolours when exposed to the air, what its smell and taste are, both when picked and after a certain time. One should also examine very carefully the hymenium, probably the most important part from the point of view of identification; as we have seen, it can be gilled, tubular, toothed, porous or smooth.

F. HOW TO DISTINGUISH BETWEEN EDIBLE AND INEDIBLE MUSHROOMS

When looking for mushrooms with a view to cooking them, there are obviously many points to consider. The leathery or woody fungi present no problem: they are simply inedible. The same applies to bitter or sharp mushrooms, unpalatable anyway; it would be very difficult indeed to chew and swallow *Lactarius rufus* or *Boletus felleus* and for this very reason they should be carefully avoided, even though not poisonous: just one specimen is enough to ruin completely a pan of delicious mushrooms.

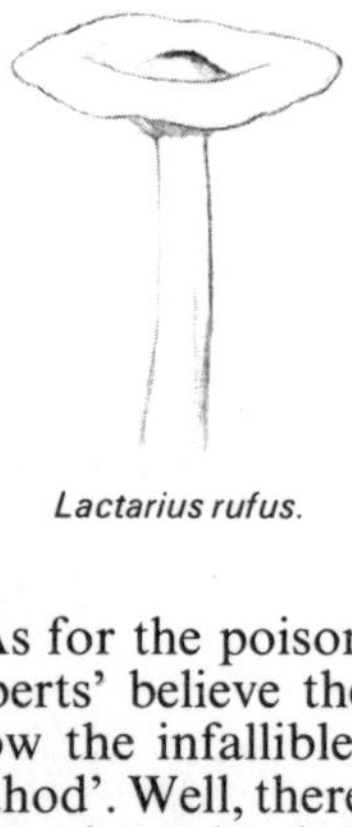

Lactarius rufus.

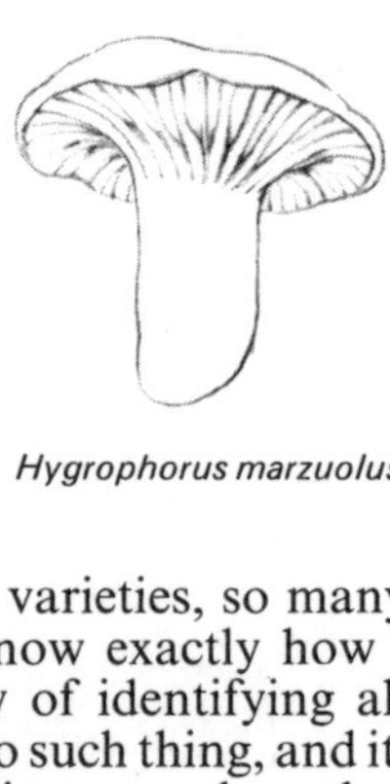

Hygrophorus marzuolus.

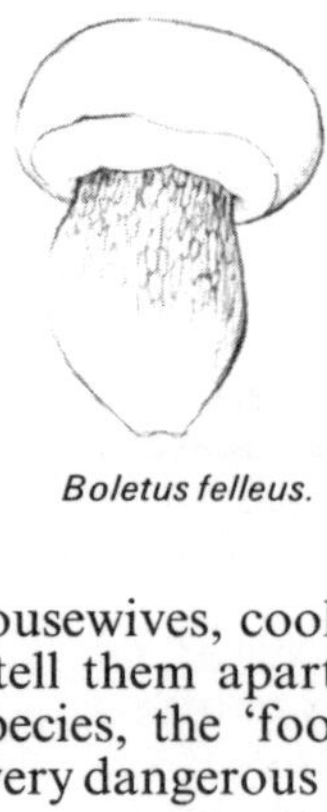

Boletus felleus.

As for the poisonous varieties, so many housewives, cooks and 'experts' believe they know exactly how to tell them apart: they know the infallible way of identifying all species, the 'foolproof method'. Well, there is no such thing, and it is very dangerous for the inexperienced to believe in any such methods.

Some mushrooms are so unpalatable as to be inedible when raw, but some of them are much improved by cooking. By the same token, not all mushrooms with a sweet and pleasant taste are edible: apparently the deadly amanites taste very good, whether raw or cooked, but just one of them can kill a few people. Their poison is not destroyed by cooking, pickling or drying. Two very different mushrooms smell like fresh flour, *Clitopilus prunulus* which is edible, and *Entoloma lividum*, which is extremely poisonous.

Some people avoid mushrooms which exude milk, believing them poisonous. It is true that some species of *Lactarius* are inedible because of their bitter taste and smell; but it is also true that *Lactarius deliciosus*, although not delicious, is good to eat and harmless, as are *L. volemus* and *L. sanguifluus*. The colour of fungi is equally useless from this point of view. It would be a mistake to avoid all purplish mushrooms, as *Rhodopaxillus nudus* is edible and even tasty when properly cooked. However, it is toxic when raw, unless it has been sliced and left for at least one hour in salt and lemon juice and the liquid thus exuded is thrown away. The same applies to other purplish or livid mushrooms, such as *Laccaria amethystina* and *Cortinarius violaceus*, which, although nothing special, are nevertheless edible and harmless. As for *Mycena pura*, it is often purplish but some experts regard it as being hallucinogenic, although not dangerously so.

Again, the discoloration of the flesh as it comes into contact with air is not a means of assessing whether a mushroom is edible or not. The whitish flesh of *Boletus satanas*, a poisonous boletus, becomes bluish when broken, reverting later to the original colour; that of *Boletus purpureus*, also poisonous, becomes first blue and then red. But the flesh of *Boletus cyanescens* becomes a very deep blue instantly; the deeper the blue, the fresher the mushroom, but also the moister the air the deeper the blue. This mushroom is not only edible but excellent. *Boletus luridus* and *B. erythropus*, also turning to blue, are poisonous when raw and edible when properly cooked. On the other hand the flesh of the deadly amanites is white when broken and remains white even after prolonged exposure to the air.

Boletus purpureus.

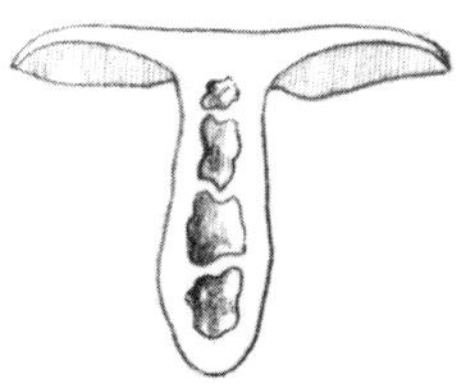

The edible *Boletus cyanescens* (section).

The colour test having failed us, we should not rely on the viscosity or otherwise of the cap. *Gomphidius viscidus, G. glutinosus, Boletus luteus, B. viscidus* and others are edible. Some mushrooms appeal to animals, particularly squirrels, and of course insects and snails or slugs; but this is no guarantee that those very mushrooms can be eaten by man as well. The deadly amanites are eaten by slugs, who thrive on them. Habitat does not have much to do with it either. *Armillaria mellea*, well known and sought after, grows on trees, which may lead someone to believe that all such fungi, unless leathery or woody, are edible. Not so: *Clitocybe olearia* and

Nematoloma fasciculare are poisonous, although the latter is so bitter nobody would dream of eating it.

To sum up, there is only one way to assess whether a mushroom is edible or not. You must be able to positively identify the individual members of common species of either kind.

G. POISONOUS MUSHROOMS

Some mushrooms are perfectly harmless when cooked but toxic when raw and can cause serious trouble according to the quantity swallowed. These are *Rhodopaxillus nudus, Boletus luridus, B. erythropus, Amanita rubescens, Amanita* (or *Amanitopsis*) *vaginata* in all its varieties, *Lepiota badhami* and almost all the species of the genera *Morchella, Helvella, Gyromitra, Peziza* and allied fungi.

Armillaria mellea is harmless when cooked and well drained, but can cause stomach troubles and nausea when raw or cooked in the oven or grilled; it should always be preboiled and the water thrown away, and even then eaten only sparingly. Other mushrooms normally considered toxic, such as *Lactarius rufus, L. necator* or *turpis, L. torminosus*, are however eaten in some countries after a careful fermentation treatment. Some others should be avoided because of their smell: *Russula foetens, Tricholoma sulphureus* and *Lepiota cristata*, which can be confused with the deadly *Lepiota helveola*.

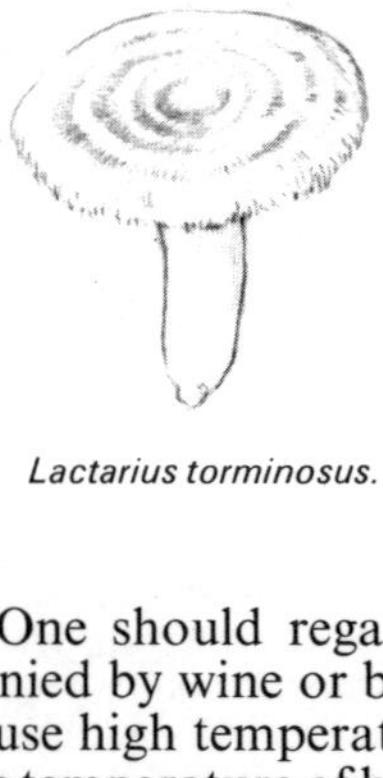

Lactarius torminosus.

Russula foetens.

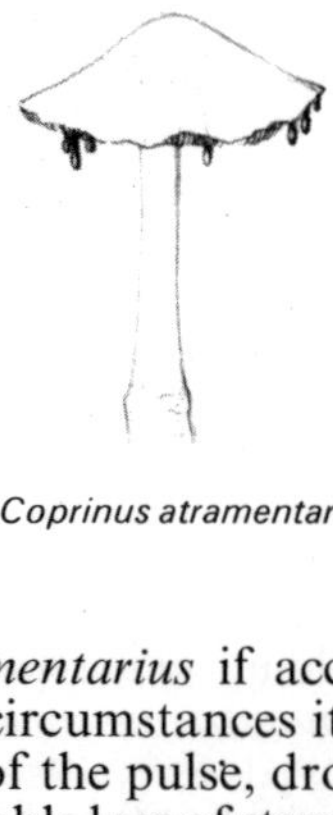

Coprinus atramentarius.

One should regard as toxic *Coprinus atramentarius* if accompanied by wine or beer, tea or coffee: in these circumstances it can cause high temperature, blushing, quickening of the pulse, drop in the temperature of hands and feet, and considerable loss of strength. These symptoms disappear within an hour or two, with the only consequence of putting you off that mushroom, but can reappear if alcohol is drunk again. Even without alcohol, this mushroom can produce the symptoms described in allergic or sensitive people and should therefore be treated with extreme care.

All the mushrooms of the *Psalliota* (or *Agaricus*) *xanthoderma* group upset the digestive system, although not seriously, whether eaten with alcohol or not. Much more serious are the consequences of eating *Psalliota infida, Clitocybe olearia, Entoloma lividum, E.*

rhodopolium, Nolanea mammosa and *N. pasqua, Tricholoma pardinum, T. virgatum* and *T. groanense, Boletus satanas, B. purpureus* and *B. lupinus,* and *Clavaria formosa.* Within an hour they cause nausea, colic, diarrhoea and sometimes fainting; the remedies consist of a strong purgative, diuretics, and the ingestion of large quantities of non-alcoholic fluids.

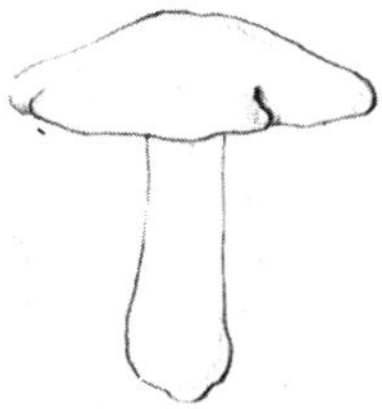

Entoloma lividum.

Amanita muscaria.

Amanita pantherina.

Inocybe patouillardii.

Clitocybe dealbata.

Other mushrooms affect the nervous system; the most common ones are *Amanita muscaria, A. pantherina, Inocybe patouillardii, Clitocybe dealbata, C. rivulosa* and *C. cerussata.* The symptoms appear within four hours of eating and consist of increased secretion of saliva, mucus and tears, weakening of the heart beat, and breathing difficulties. Here again one should administer a strong purgative, diuretics and non-alcoholic drinks; if necessary, a doctor should inject with atropine.

Amanita muscaria should also be avoided, even when its poison is removed. Even after years of safely eating it, serious poisoning can still occur. It is known that some primitive people dry this mushroom and derive from it a stimulating drug—with what results it is not certain.

The nervous system is also damaged by the beautiful *Russula emetica*; fortunately, however, it is an almost immediate emetic and, being quickly expelled, stands little chance of causing serious damage. It is also rather sharp, which makes it difficult to eat in quantity, and it is said that one specimen only, cooked with other good mushrooms, acts as a condiment without causing harm. It is not, however, to be recommended.

Russula emetica.

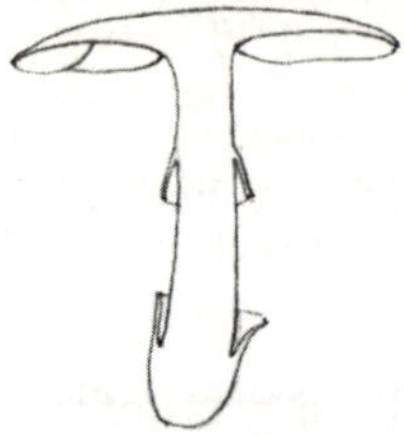

The deadly *Amanita phalloides* (section).

Some species are much more dangerous than any of the above. One of them is extremely common in any habitat, be it mountain or plain, frondose or coniferous woods, from spring to late autumn: this is *Amanita phalloides*, in its forms *viridis* and *virescens*. Twenty grams of raw mushroom are enough to kill an adult; and the poison persists through cooking, preboiling, or drying, as it does in the case of the other three deadly amanites: *A. verna*, rare in spring, *virosa*, growing from spring to autumn, and *bisporigena*, growing in the oak woods of Northern America. Together with *Lepiota helveola, Cortinarius orellanus* and *Gyromitra esculenta*, these mushrooms attack liver and kidneys and are a serious threat to life. The case of *Gyromitra esculenta* is a rather peculiar one. Its name means that it is edible and it was given to it because it has been eaten since the dawn of time by millions without any side-effect. However, it has been associated with some serious, and even lethal, cases of poisoning. Why and how it is not clear. Most of its toxins are destroyed by cooking or drying; and some people can eat it even when raw, but their example is not to be followed. It should be carefully cooked, its juices drained off, and should be eaten in small quantities (although it does taste excellent), leaving an interval of a few days before eating it again.

One of the most serious cases of poisoning is the one produced by *Cortinarius orellanus*, as the symptoms can appear up to two weeks after eating the mushroom, by which time it is easy to overlook the connection. In contrast, the symptoms caused by one of the amanites begin within eight and forty-eight hours; after a short pause during the first two days, they come back with deadly inexorability to end only with the death of the patient after two or three weeks of agonising suffering. The only help the layman can offer to anyone poisoned by amanites, *Lepiota helveola, Cortinarius orellanus* and *Gyromitra esculenta*, is to rush the patient to the nearest hospital. There doctors will try to fight the toxins which attack the liver, avoid the dehydration of the organism, help the kidneys, support the nervous system and the heart and alleviate the pain. While waiting for medical help, the patient should always, whatever the poisoning, be kept warm; he should be given a purgative which, in the case of strong abdominal pain, should be an oily one, and several enemas of olive or corn oil mixed with warm water and 10–20 drops of laudanum. One hour after the purgative he should be given some hot vegetable stock, or some milk, and should be encouraged to vomit. If he does so too violently, he should be given a cold fizzy drink or some ice; if he can not do it at all, it means that the poison has already progressed into the intestines. Any surplus from the suspect meal should be kept, as well as fragments of food he might have vomited, as their examination would aid the identification of the cause of the poisoning. If it is known that the patient has eaten deadly mushrooms, he should be given a glass of water containing a teasponful of salt alternated every hour with a glass of water containing the same amount of sugar.

But to end on a happy note: the Max Planck Institute at Munich is looking into the possibility of using the toxins of *Amanita phalloides* in the fight against cancer. And let's remember that without fungi we would not have antibiotics, wine, beer, yeasts and some cheeses.

H. NUTRITIONAL VALUE OF MUSHROOMS

Do mushrooms have any nutritional value? The alimentary substances of our foodstuffs are either plastic, ie constitute the organic tissues; energetic, ie essential to organic functions; or plastico-energetic. Protids, like meat, belong to the first kind; glucids, like bread, to the second; and lipids, like butter, to the third. In order to be digested properly, they have to be ingested with water and cellulose; the latter is not absorbed by the organism but helps in softening the foodstuffs and renders them pliable to the action of the intestines. The digestive process also needs condiments and vitamins; the former, like salt, stimulate the gastric fluids; and the latter act as catalysts to the biochemical process.

Some mushrooms are undoubtedly excellent condiments; they also contain several vitamins. *Catharellus cibarius* and *Amanita caesarea* contain a proportion of Vitamin A; the vitamins of the B group, including PP, are found in many species as long as they are not dried; Vitamin C abounds in *Fistulina hepatica*, but not so much in any other fungus; and many other species contain Vitamins D2 and K. All edible mushrooms contain plenty of cellulose as well as water and, in various forms, useful substances such as phosphorous, potash, zinc, copper, iron and others. As for the basic alimentary substances, they contain up to 10% of their weight of glucids, 5% of protids, and just about 1% of lipids. According to the scales of the analysts, their composition is therefore close to that of meat and eggs, much richer, that is, than other vegetables. It all depends on how much of this richness can actually be used by the healthy human organism. Sometimes mushrooms are difficult to digest and sometimes they go right through the digestive system without affecting it or being affected by it. At any rate, they are definitely among the best of condiments.

I. MYCOLOGISTS AND MYCOLOGY

Mycology is rather a new science. Some fungi have been eaten since antiquity, and others avoided as being poisonous; but as far as we know, only the Greek physician Hippocrates (460–377BC), knew them well enough to use against certain afflictions. Which fungi he used, against which afflictions and how, unfortunately we do not know. Theofrastus, a Greek botanist (372–287BC), belied the current opinion of his day, which regarded mushrooms as plants' pathological tumours, and declared them to be plants in their own right, albeit fruitless. He did not go so far as to discover they were the actual fruit of plants living under the soil. Pliny the Elder, the Roman naturalist (23–79AD), discusses fungi at length in his *Naturalis historiae libri XXXVII*, and even gives rules to distinguish the edible from the poisonous ones. The Greek doctor Dioscorides (1st century AD) gave instructions on how to treat people poisoned by mushrooms—or even those who were suffering from indigestion from having eaten too many—by administering salt water emetics. The Persian doctor Avicenna (980–1037) states that all green fungi are poisonous, which is not true, although the deadly *Amanita phalloides* is greenish in some of its forms. Saint Albert the Great of Cologne (c1200–1280), an encyclopedic genius, gives advice on how to fight the poison of *Amanita muscaria* and how to use such a mushroom to kill flies.

A few centuries later the Frenchman Charles de l'Ecluse (1525–1609) first described and illustrated about one hundred species of fungi. Mycology was greatly improved by the Swede Carl von Linne (Linnaeus, 1707–1778), the author of more than 200 works on systematic and descriptive botany; we owe to him the introduction of the binomial nomenclature—the practice of classifying a plant by quoting both its genus and its species. The first to cultivate mushrooms was the Florentine Pier Antonio Micheli (1679–1737), who used the spores. But the universally-recognised founder of modern mycology is the Dutch Christian Persoon (1755–1837), who, in his *Synopsis methodica fungorum*, improved both the scientific classification of fungi and the methods used to determine their genus and species. Another step forward is represented by the work of Elias Fries (1794–1878), a Swede, who classified fungi according to the colour of their spores and first formulated those mycological principles which are still valid and accepted today.

Among the other great mycologists one should mention the Frenchmen L. Quélet (1832–1899) and N. Patouillard (1854–1926); the Italians Pier Andrea Stoccardo (1845–1920), the author of *Sylloge fungorum omnium hucusque cognitorum*, and Monsignor Giacomo Bresadola (1847–1929), who wrote, among other works, an *Iconographia Mycologica* in twenty-four volumes, which is still being supplemented by other authors.

J. FURTHER READING

AINSWORTH, G. C. & SUSSMAN, A. S. (Ed), *Fungi: An Advanced Treatise*, 5 vols, Academic Press (1965–74).
AINSWORTH, G. C. and BISBY, G. R., *Dictionary of the Fungi*, CMI, Kew (1961).
ALEXOPOULOUS, C. J., *Introductory Mycology*, John Wiley, New York (1963).
BOURDOT, H. and GALZIN, A., *Hymenomycetes de France*, Societé Mycologique de France, Paris (1927).
BRESADOLA, G., *Iconographis mycologica*, 24 vols, Milan (1927–38).
COOKE, M. C., *Illustrations of British Fungi*, London (1880–90).
CORNER, E. J. H., *A Monograph of Cantharelloid Fungi*, OUP (1966).
CORNER, E. J. H., *A Monograph of Clavaria and Allied Genera*, OUP (1950).
DANCE and BYARD, *Field Guide to Mushrooms and Toadstools*, Collins (1974).
DENNIS, R. W. G., 'The Genus Inocybe' *The Naturalist*, London (1954).
DENNIS, R. W. G., 'The Genus Mycena' *The Naturalist*, London (1954).
DENNIS, R. W. G., *British Cup Fungi and their Allies*, Ray Society, London (1960).
DENNIS, R. W. G., *British Ascomycetes*, J. Cramer, Braunschweig (2nd ed, 1978).
DENNIS, R. W. G., ORTON, P. D. and HORA, F. B., 'New Check List of British Agarics and Boleti', *Suppl. Trans. Brit. Myc. Co.* (1960).
DICKINSON & LUCAS (Ed), *Encyclopaedia of Mushrooms*, Orbis (1979).
DRING, D. M. (Ed), *Mushrooms and Fungi* (1978).
ELLIS, A. E., *British Fungi*, Jarrold (1976).
FINDLAY, W. P. K., *Wayside and Woodland Fungi*, Warne (1967).
FINDLAY, W. P. K., *The Observer's Book of Mushrooms, Toadstools and other Fungi*, Warne (1978).
GRIGSON, J., *Mushroom Feast*, Michael Joseph (1975).
HAAS & GOSSNER, *Fungi*, Burke (1969).
HEIM, R., *Champignons d'Europe*, N. Boubée, Paris (2nd ed, 1969),
HIGGINS, V. (Ed), *Mushrooms and Toadstools in Colour*, London (1961).
HORA, F. B., 'The Genus Panaeolus in Britain', *The Naturalist* (1957).
HVASS, E. & H., *Mushrooms and Toadstools in Colour*, London (1961).
JORDAN, M., *Guide to Mushrooms*, Millington (1975).
KIBBY, G., Mushrooms and Toadstools: A Field Guide, OUP (1979).
KÜHNER, R. and ROMAGNESI, H., *Flora Analytique des Champignons Superieurs*, Masson et Cie, Paris (1953).
LANGE, M. and HORA, F. B., *Collins Guide to Mushrooms and Toadstools*, Collins (1963).
LISTER, A. and G., *A Monograph of the Mycetozoa*, British Museum (3rd ed, 1925).
LOEWENFELD, CLAIRE, *Britain's Wild Larder: Fungi*, Faber & Faber (1956).
MASSEE, G., *British Fungus Flora*, Vol I–IV, London (1892–3).
MAUBLANC, A., *Les Champignons de France* I and II, P. Lechevalier, Paris (1946).
NICHOLSON, B. E. & BRIGHTMAN, F. H., *Oxford Book of Flowerless Plants*, Oxford (1966).
ORTON, P. D., 'Cortinarius', *The Naturalist* (Supplement, 1955).
ORTON, P. D., 'New Check List of British Agarics and Boleti: Part III, Notes on Genera and Species in the List', *Trans. Brit. Myc. Soc.* 43 (2), 159–468 (1960).
PEARSON, A. A., 'British Boleti', *The Naturalist*, London (1946).
PEARSON, A. A., 'The Genus Lactarius', *The Naturalist*, London (1950).
PEARSON, A. A., 'The Genus Russula', *The Naturalist*, London (1948).

PEARSON, A. A., 'The Genus Mycena', *The Naturalist*, London (1955).
PEARSON, A. A., 'The Genus Inocybe', *The Naturalist*, London (1954).
PILAT, A. & USAK, O., *Mushrooms and other Fungi*, London (1961).
RAMSBOTTOM, J., *A Handbook of The Larger British Fungi*, British Museum (Natural History), London (1923).
RAMSBOTTOM, J., *Edible Fungi*, London (1943).
RAMSBOTTOM, J., *Handbook of Larger Fungi*, London (1951).
RAMSBOTTOM, J., *Mushrooms and Toadstools*, Collins (1963).
RAMSBOTTOM, J., *Poisonous Fungi*, King Penguin (1945).
REA, C., *British Basidiomycetae*, Cambridge (1922).
RINALDI, A. & TYNDALA, V., *Mushrooms and other Fungi*, Hamlyn (1972–4).
ROMAGNESI, H., *Petit Atlas de Champignons*, 3 vols, Bordas, Paris (1962).
SOOTHILL & FAIRHURST, *The New Field Guide to Fungi*, Michael Joseph (1978).
TOSCO, U., *Mushrooms in the Wild*, Orbis (1977).
TOSCO, U. & FANELLI, A., *Mushrooms and Toadstools: How to Find and Identify Them*, Orbis (1976).
WAKEFIELD, E., *Observer's Book of Common Fungi*, London (1954).
WAKEFIELD, E. M. and DENNIS, R. W. G., *Common British Fungi*, Gawthorn, London (1950).
WATLING, R. *The Naturalist Keys to Agaric Genera*, Available from the Editor of the Naturalist, University of Bradford (1978).
WATLING, R., *The Identification of the Larger Fungi*, Amersham (1973).
WATLING, R., *Mushrooms and Toadstools of Broadleaved Forests*, Forestry Commission Record 106, HMSO (1975).
WATLING, R., *Mushrooms and Toadstools of Coniferous Forests*, Forest Commission Record 107, HMSO (1976).
WATLING, ROY, *British Fungus Flora* (Agarics and Boleti), HMSO, Edinburgh (1969).
WEBSTER, J., *Introduction to Fungi*, CUP (2nd ed, 1980).

HOW TO READ THE SYMBOLS

Months during which carpophores are produced

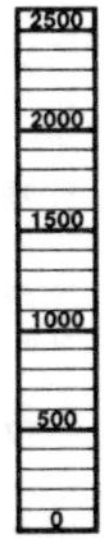

Maximum altitude of the habitat

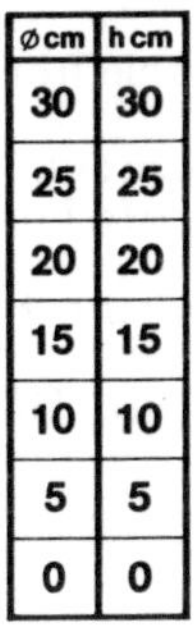

Dimensions of the carpophore:
diameter and height in cm

EDIBILITY

 1 Edible

 2 Inedible as suspect or unsavoury

 3 Poisonous

TYPE OF FUNGUS

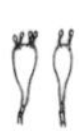 4 Basidiomycete

 5 Ascomycete

 6 Gymnocarpous

 7 Angiocarpous

 8 Hemiangiocarpous

COLOUR OF SPORES

A 9 (*alba*) white

R 10 (*rosea*) pink

O 11 (*ochracea*) ochre

B 12 (*brunnea*) brown

V 13 (*violacea*) purplish

N 14 (*nigra*) black

HABITAT

15 Underground

16 On trunks, boles and woody remains

17 Under frondose trees, ie oaks, beeches, etc

18 Under conifers, ie larches, firs, pines, etc

19 Under shrubs

20 Meadows

21 Sandy soils

22 Seaside

HYMENIUM

23 Smooth

24 Toothed

25 Porous

26 Tubular

27 Gilled

28 Pseudo-gilled, or wrinkled

CAP

29 Concave

30 Convex

31 Umbonate with a central boss

32 Campanulate, bell-shaped or conical

33 Gibbous (ie with humps)

34 Warty (ie covered by fragments of the general veil)

35 Squamose or fibrillose, scaly or covered with silk-like fibres

36 Spiky

37 Scaly

38 Areolate (ie the skin is fragmented into more or less regular portions)

39 Velvety

40 Viscid or glutinous, particularly in wet weather

41 Zonal (ie with more or less concentric stripes of different colours)

42 Hygrophanous (changing colour, usually to a deeper shade, in wet weather)

43 Easily peeled

44 Fleshy

45 Thin

CAP MARGIN

46 Involute, turned inwards

47 Fringed by the remnants of the general or partial veil

 48 Overhanging or jutting

 49 Radially striate

 50 Radially furrowed

 51 Wavy or lobed

GILLS

 52 Shallow

 53 Deep

 54 Crowded and, generally, thin

 55 Distant and, generally, thick

 56 Forked, either with the tip subdivided or joined by transverse veins

 57 Marbled, of uneven colour

 58 Adnexed (just reaching the stipe)

 59 Free from the stipe

 60 Adnate (broadly adhering to the stipe)

 61 Decurrent (prolonged down the stem)

 62 Briefly decurrent by means of a fillet or hook

 63 Emarginate (having a notch on the end)

 64 Serrated (on the edge)

 65 Irregularly dented (on the edge)

TUBES

 66 Decurrent

 67 Decurrent with fillet

 68 Free

 69 Adnate

STIPE

 70 Absent, sessile (closely attached without a stem) fungus

 71 Clavate or club-shaped (thickened upwards)

 72 Pear-shaped

 73 Narrowing at the top

 74 Widening at the top

 75 Wavy or bending

 76 Eccentric

 77 Lateral

 78 Marked by annular (ring-like) areas of different colour

 79 Scrobiculated, or pock-marked

 80 Fistulous, empty inside like a small tube

 81 Tubular, empty like a large tube

 82 Cavernous, with one or more hollows

 83 Reticulated, with a fine lattice-work pattern

 84 Heterogeneous (different tissues from those of the cap and easily separated)

 85 Homogeneous (same tissue as the cap, can only be broken off)

 86 Sheathed to the base of the ring

RING

 87 Drooping

 88 Lifted at the margin

 89 False ring, remnant of the cortina

CORTINA

 90 Stringy or glutinous cortina

FOOT

 91 Swollen

 92 Bulbous

 93 Rooting

 94 With mycelium fragments

 95 Sheathed in mycelium cotton

VOLVA

 96 Membraneous

 97 Truncated, with a smooth brim

 98 Sheathing, tall and adhering to the stipe

 99 Dissociated, breaking up into scales

MILK

 100 A fungus exuding milk on being broken or incised

PLATES

The species have been grouped according to their dominant colour—red, white, grey, orange, green, off-white, violet, yellow, fawn and brown. For ease of reference, these colours are indicated in the top corner of each page.

Caesar's Amanita or Caesar's Mushroom

Cap globular, becoming less convex, orange, later with striate margin. **Gills** yellow. **Stipe** becoming tubular, yellow, with yellow ring and white, ample volva of good consistency. **Flesh** tender and white, yellowish under the skin of the cap and the outer wall of the stipe. **Smell** faint. **Taste** pleasant. **Spores** white touched with yellow, 9–13 × 6–7·5 μ; elliptical-oblong. **Habitat** under frondose trees in the Mediterranean basin, becoming rare towards the north. **Edible/Poisonous** edible; of excellent quality whether raw or cooked.

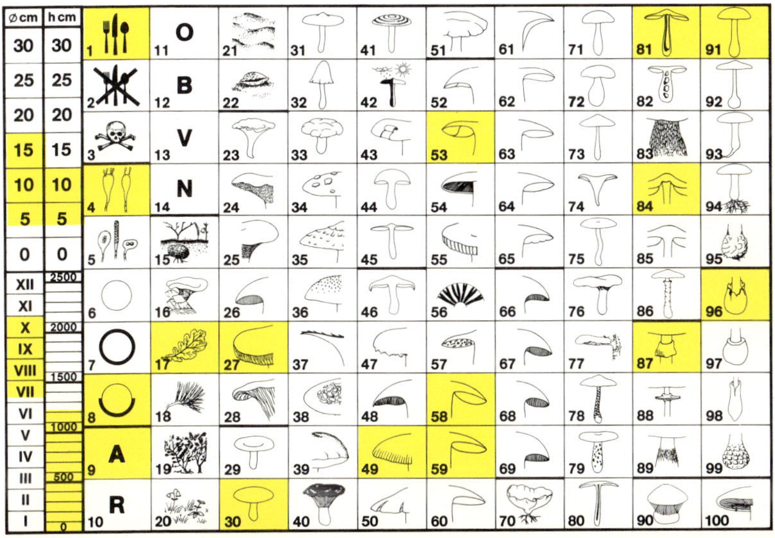

Fly Agaric

Cap globular then plane; red, viscid in wet weather, covered in white soft warts easily removed by the rain; margin becoming striate; the variety *aureola* has a margin precociously striate and generally no warts. **Gills** white. **Stipe** soon fistulous, then tubular, white; white striate ring with yellow margin. **Flesh** white, yellowish under the cuticle of the cap. **Smell** almost imperceptible. **Taste** nutty. **Spores** white, 9·5–12 × 6–8 µ; elliptical-oblong. **Habitat** already growing in summer but usually in the autumn, in tufts, very common. **Edible/Poisonous** poisonous, but can be eaten after certain precautionary treatment.

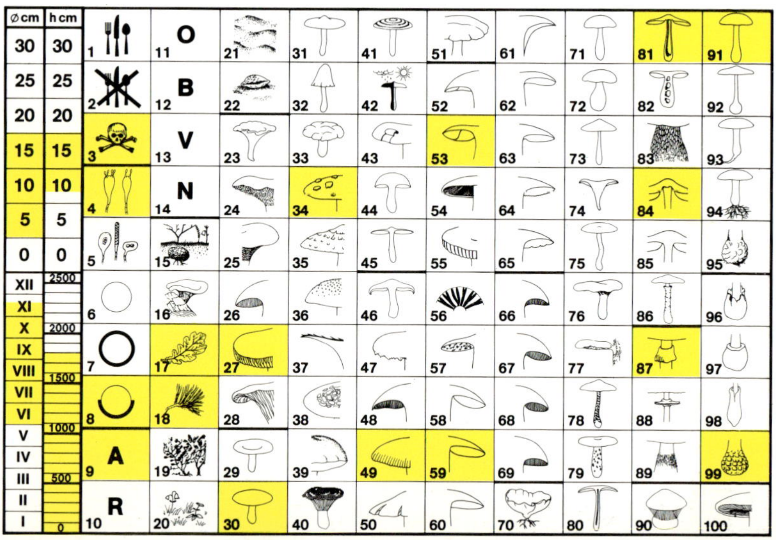

Cortinarius cinnabarinus Fries

Red cortinarius or Cinnabar cortinarius
Syn. *Dermocybe cinnabarina* (Fr.) Wünsche
Cap dry and shiny, scarlet red, becoming orange with age. **Gills** concolourous with the cap, becoming much darker with age. **Stipe** concolourous with the cap, with fine remnants of the cortina. **Flesh** pink in the cap, reddish in the stipe. **Smell** imperceptible. **Taste** imperceptible. **Spores** fawn-brown, 8–10 × 5–6 μ; amygdaliform. **Habitat** on siliceous soils. **Edible/Poisonous** not to be taken, although not poisonous, as is easily mistaken for *Cortinarius orellanus* which is deadly.

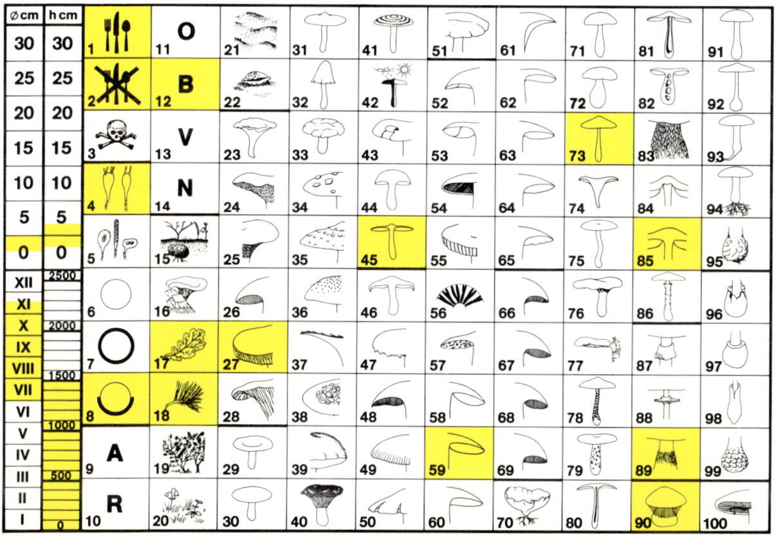

Russula lepida Fries

Rose russula
Syn. *R. rosacea* Pers. ex S. F. Gray
Cap maintains an involute margin even when it becomes plane or concave; dry, hard to peel; red, sometimes with yellow spots; rarely totally white. **Gills** white, at times with a red margin or tip; fragile. **Flesh** firm in the cap, stringy in the stipe. **Smell** of cedar wood. **Taste** bitter after some time. **Spores** white tinged with yellow, 8–9 × 7–8 μ; globular-oval. **Habitat** from late spring to early autumn, under broad-leafed trees, rarely under conifers; alone or in small groups. **Edible/Poisonous** not to be taken although edible, because of poor quality and rather bitter even after careful cooking.

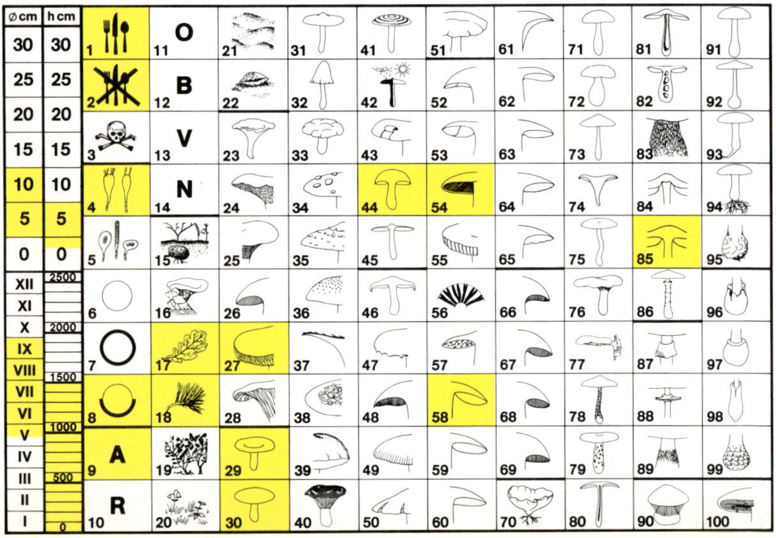

The Sickener

Cap first hemispheric, then plane or slightly depressed, with a striate border; red or pinkish, sometimes with yellowish spots; during persistent rain it discolours to become whitish with only a few spots of pink. **Gills** extremely fragile, free or very slightly joined to the stipe; white, then tinged with cream. **Stipe** wrinkled, fragile, white, first full then with a marrow. **Flesh** white, but slightly pink under the cap's cuticle. **Smell** of fruit. **Taste** sharp almost immediately. **Spores** white, 8–11 × 8–10 μ globular with spikes and apex. **Habitat** mainly under chestnuts and oaks, in groups or scattered. **Edible/Poisonous** poisonous.

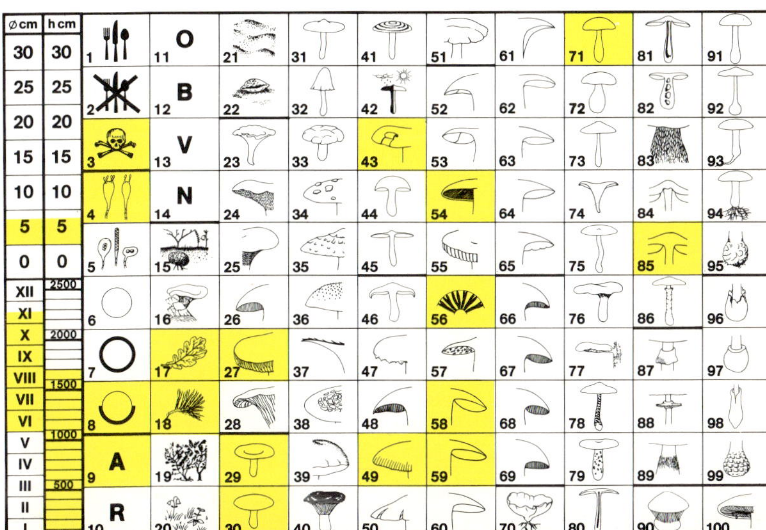

Beefsteak Fungus

Cap semicircular or tongue-shaped, dark red tinged with pink; often with dark spots along the margin; glutinous, velvety. **Tubes** short, separated like the bristles of a brush; reddish. **Pores** small, round, yellowish then reddish becoming brownish when bruised. **Stipe** sometimes absent, usually short and stocky, the colour of the cap; solid. **Flesh** firm and succulent, red with pale veins; glutinous at the end. **Smell** mild. **Taste** slightly sharp. **Spores** pale pink, 4–5·5 × 3–4 μ; oval. **Habitat** the trunks of old oaks and chestnuts. **Edible/Poisonous** edible; good quality even when raw, but in this case one should drain off the liquids exuded when dressed.

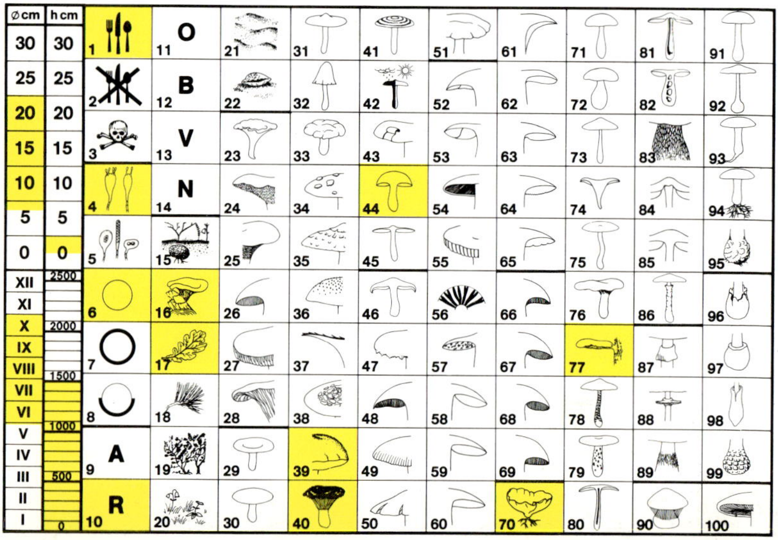

Scarlet Elf Cup

Similar to a small cup, sometimes without stipe, thin and fragile, no smell and no taste; the internal surface is scarlet, paling towards the margin; the outer surface also red, rarely whitish, but appearing to be whitish or ash-like or purplish owing to the bloom which covers it. **Spores** white, 25–35 × 10–13 µ; elliptical-oblong. **Habitat** at the end of winter on the branches of conifers, rarely of broad-leafed trees. **Edible/Poisonous** edible even when raw.

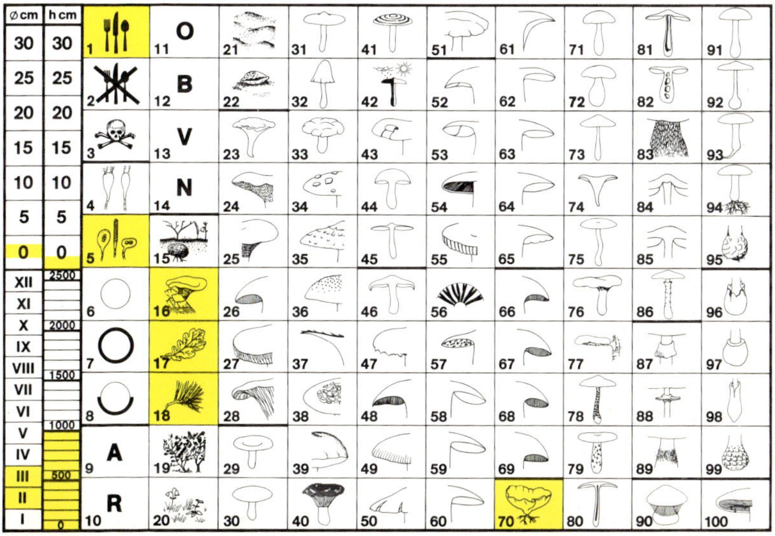

Syn. *Amanita strobiliformis* (Vittadini) Quélet
Cap globular, then expanded and finally plane; white, shiny; few wide warts, hairy and ash-like, then ochre and fugacious; fringed margin. **Gills** white then creamy. **Stipe** white covered in mealy specks or scales, creamy and non-persistent; wide and mealy ring; thick foot, with soft remnants of the volva. **Flesh** white. **Smell** almost imperceptible. **Taste** pleasant. **Spores** white, 10–13 × 7–8 μ; oval. **Habitat** in the woods, rarely outside them, alone or in small groups. **Edible/Poisonous** edible with delicate flavour.

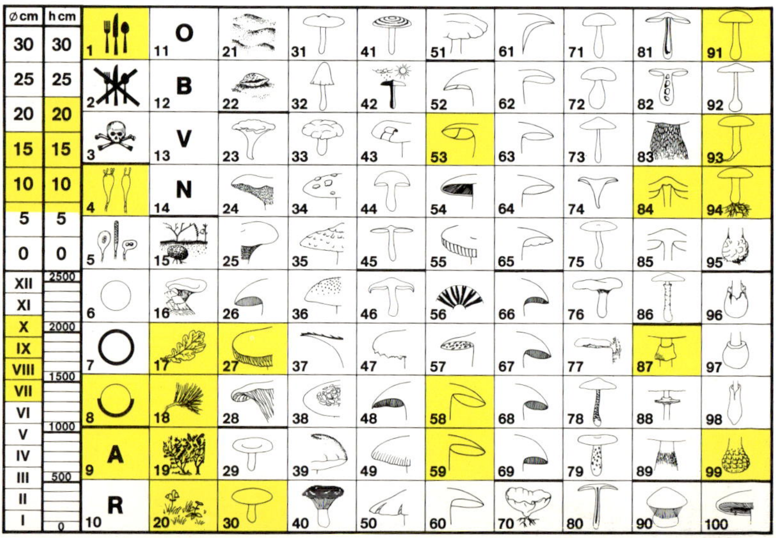

Cap globular, then expanded; moist, smooth and white. **Gills** white, with irregular edge. **Stipe** solid, white and smooth, but slightly flaky above the ring; ring creamy white and fugacious; volva wide and membraneous, white or ochre. **Flesh** white and tender. **Smell** mild. **Taste** pleasant. **Spores** white, 9–12 × 6–7·5 μ; elliptical. **Habitat** also along the sea shore, in small groups usually quite tight together. **Edible/Poisonous** edible but mediocre.

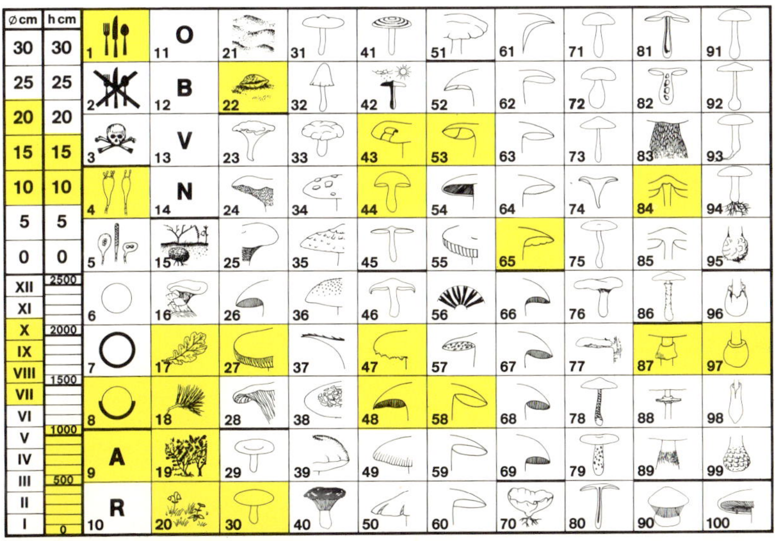

Fool's Mushroom or Spring Amanita

Cap globular or hemispherical, then expanded; white, or tinged with pinkish-ochre on the top. **Gills** white, then creamy. **Stipe** slender, as white as the ring and volva; becoming pinkish-ochre when bruised. **Flesh** tender, white. **Smell** imperceptible. **Taste** acid after some chewing (do not swallow the fragment being chewed!). **Spores** white, 8–10 × 7–9 μ; almost globular. **Habitat** particularly under oaks, in spring, sometimes again in the autumn. **Edible/Poisonous** deadly like *Amanita phalloides*.

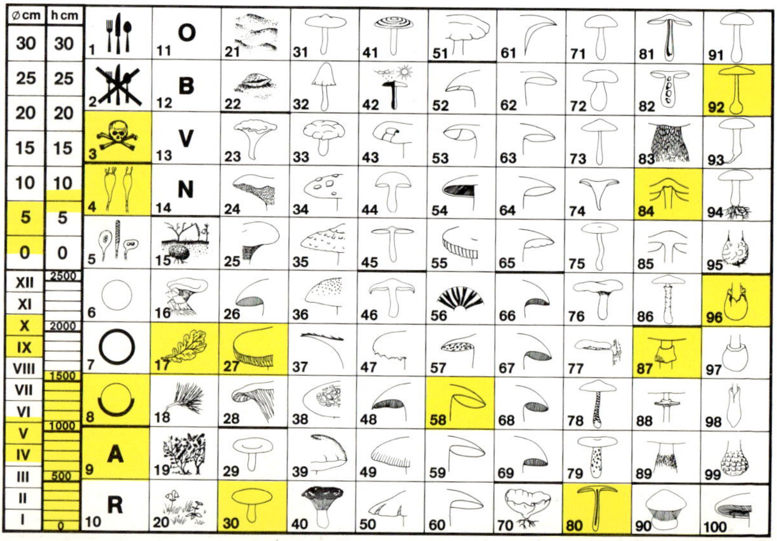

Destroying Angel
Cap globular then conic and finally expanded, maintaining a protruding summit; white or tinged with ochrish-pink, fibrillose and silky. **Gills** white. **Stipe** white, covered in fugacious woolly flakes; ring thin, white, deformed, fugacious; bulbous foot, white volva, persistent. **Flesh** tender, white. **Smell** imperceptible or slightly unpleasant. **Taste** slight but unpleasant (not to be swallowed!). **Spores** white, 9–12 μ; globular. **Habitat** mainly under beeches; rare in spring, rare under conifers; not widely spread. **Edible/Poisonous** deadly like *Amanita phalloides*.

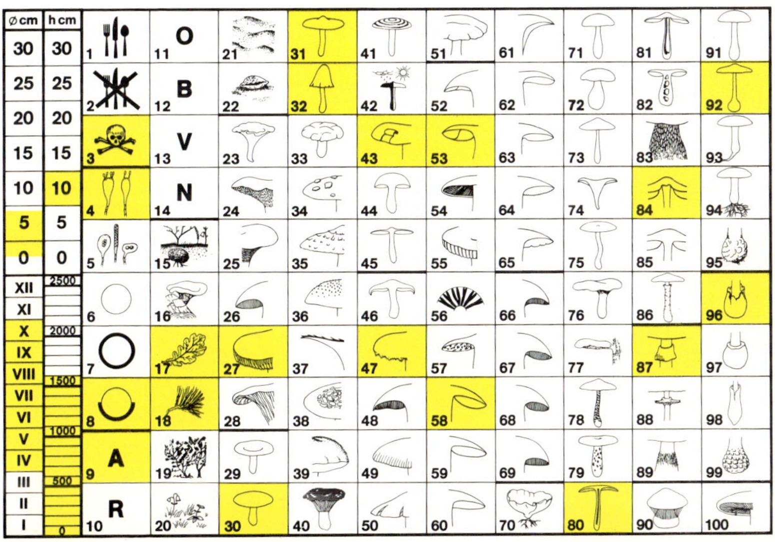

Smooth Lepiota or Nutshell Lepiota
Cap white, sometimes tinged with ash or ochre, particularly in the centre; occasionally moderately umbonate; often with slightly festooned margin. **Gills** white, then pink. **Stipe** generally with bulbous foot; white, with thin, fugacious ring. **Flesh** fragile and white. **Smell** pleasant. **Taste** sweetish. **Spores** pink, 7·5–9·5 × 5–6 μ; oval. **Habitat** in fields and gardens in spring, late summer and autumn. **Edible/Poisonous** edible; of good quality.

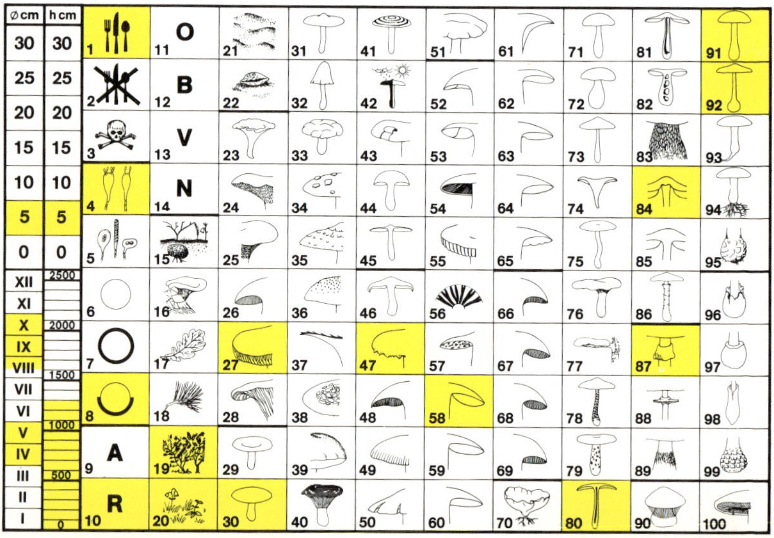

Horse Mushroom
Cap first globular or hemispherical, white, smooth; then flattening or even concave, tinged with yellow; becoming yellowish when bruised. **Gills** white, then pink and finally deep brown or blackish. **Stipe** white, yellow where bruised; ring white, soft, wide, double, the inferior edge like a toothed wheel. **Flesh** firm, white or yellowish. **Smell** mild, of aniseed. **Taste** mild, of aniseed. **Spores** reddish-brown, 7–8 × 4·4–5·5 μ; oval. **Habitat** on open and sunny soils. **Edible/Poisonous** edible; of good quality.

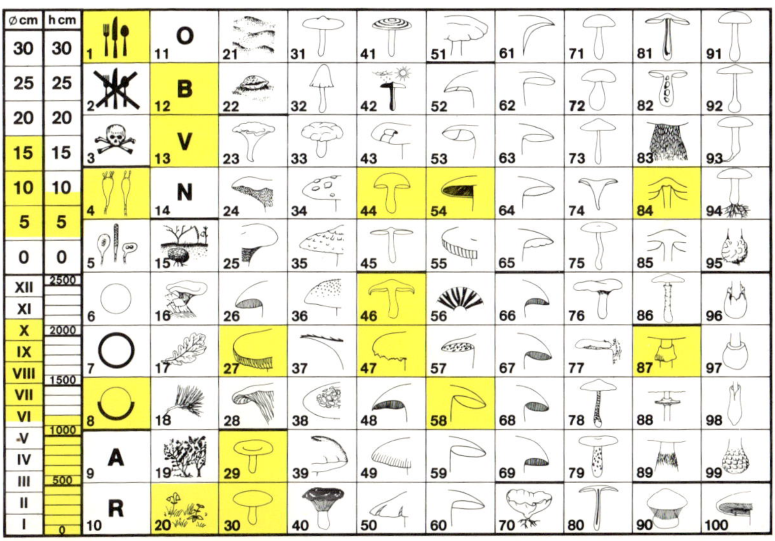

Slimy Beech Tuft or Porcelain Fungus
Syn. *Oudemansiella mucida* (Scrad. ex Fr.) v. Hoehn.
Cap with thin margin, transparent so as to allow the gills to be seen; mucilaginous, shiny, white, sometimes tinged with ash or olive. **Gills** distant, deep, occasionally slightly decurrent; white. **Stipe** fibrous and white. **Flesh** gelatinous, springy, white. **Smell** faint. **Taste** faint. **Spores** white, 15–18 μ; globular. **Habitat** on tree trunks, at a certain height, in tufts; rare. **Edible/Poisonous** edible.

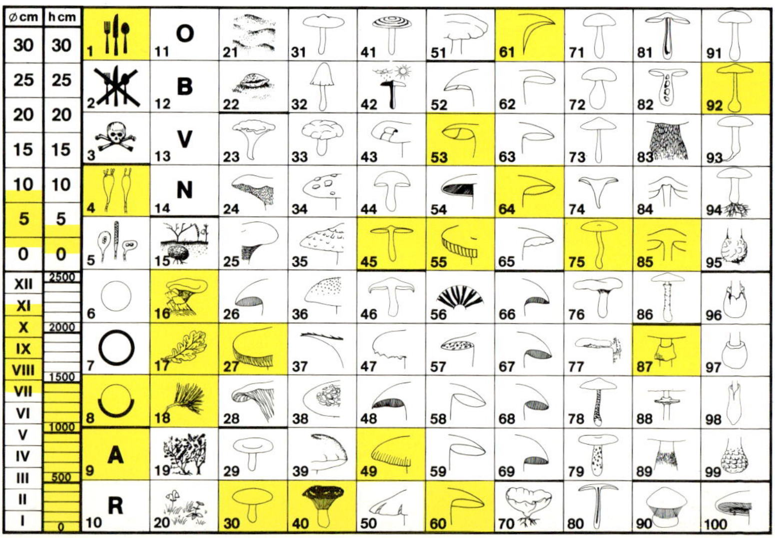

Rust-spot Fungus or Spotted Tough Shank
Cap white, silky, tinged with nut-brown, with occasional rusty spots; thin and smooth edge. **Gills** white, then yellowish with rusty spots. **Stipe** often slightly crooked, with thin or root-like foot; white and spotted like the cap. **Flesh** firm in the cap, fibrous in the stipe; white. **Smell** slightly nauseous. **Taste** somewhat bitter. **Spores** white or tinged with yellow, 4–6 × 3–5 μ; elliptical-globular. **Habitat** both on the ground and on the boles of trees, live or dead; in groups or tufts. **Edible/Poisonous** inedible, as it is bitter even when properly cooked.

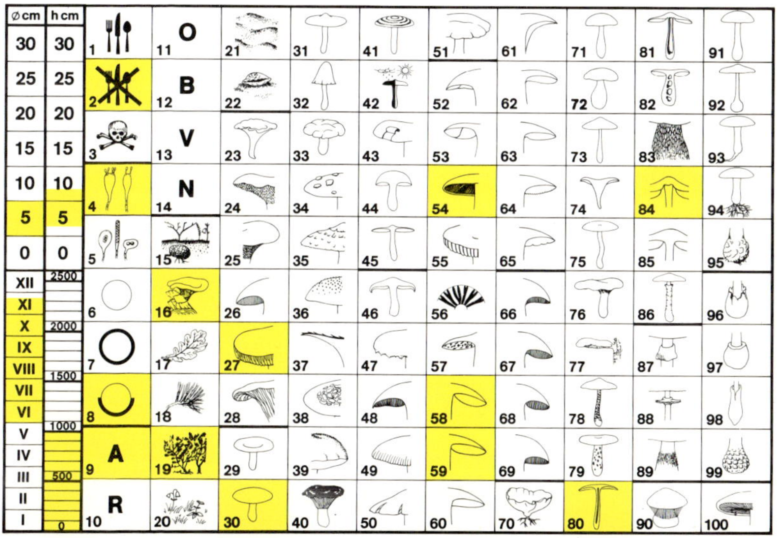

Dove-coloured Tricholoma
Cap with often wavy margin; white, silky, shiny, with some pink or green or yellow spots in old specimens. **Gills** with toothed or indented edge; white then creamy. **Stipe** slightly swollen at the foot; white, silky, sometimes greenish at the foot; solid. **Flesh** firm in the cap, fibrous in the stipe, white. **Smell** faint, of flour. **Taste** mild, of flour. **Spores** white, 5·5–7 × 4–5 μ; elliptical or oval. **Habitat** under frondose trees in groups. **Edible/Poisonous** edible; of good quality.

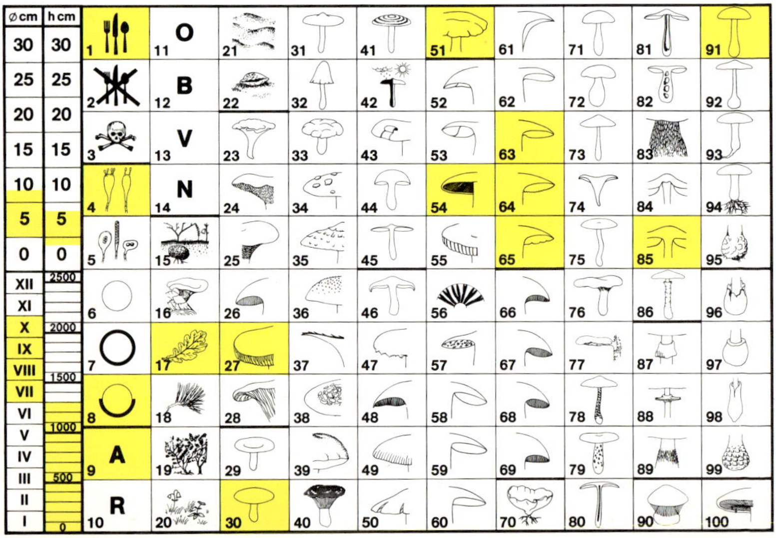

White-lead Clitocybe
Cap occasionally umbonate, sometimes concave; often with lobed margin; white, smooth, with the occasional yellowish spot. **Gills** very crowded, white then creamy. **Stipe** white and downy near the cap, whitish and flaky in the lower part; somewhat swollen. **Flesh** fairly firm, white. **Smell** of flour. **Taste** of flour, but not pleasant (do not swallow). **Spores** white, 4–5 × 3–4 μ; globular or oval. **Habitat** in the woods, both on low and high ground, sometimes in circular groups. **Edible/Poisonous** poisonous.

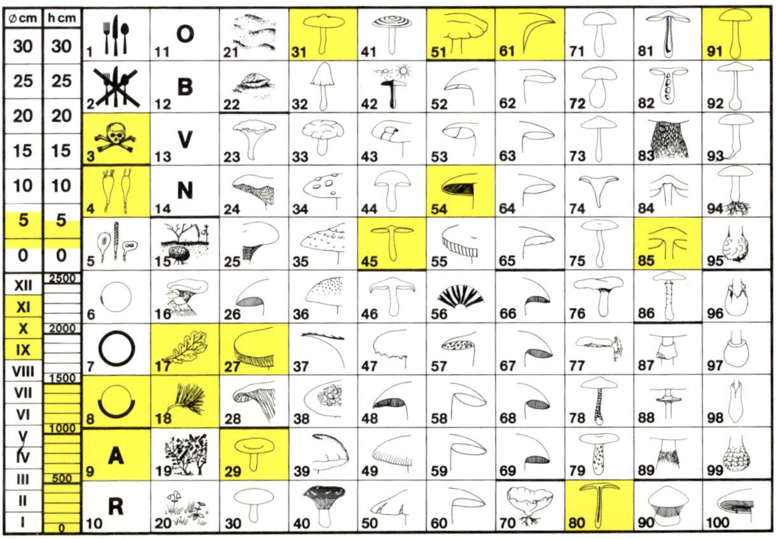

18 Leucopaxillus paradoxus (Costantin et Dufour) Boursier

White Paxillus

Cap very fleshy, sometimes umbonate, drooping and often striate margin; dry, quite often cracked and scaly; white, then faded yellow. **Gills** somewhat distant, occasionally cross-joined; slightly decurrent; white then creamy. **Stipe** pear-shaped, sometimes bent, often scaly; white or tinged with cream; full. **Flesh** firm and white. **Smell** pleasant. **Taste** mild. **Spores** white, 6–8 × 4–5·5 μ; elliptical, with tiny spines. **Habitat** in summer under conifers, sometimes also under oaks; in groups or in tufts of several specimens joined at the foot. **Edible/Poisonous** edible.

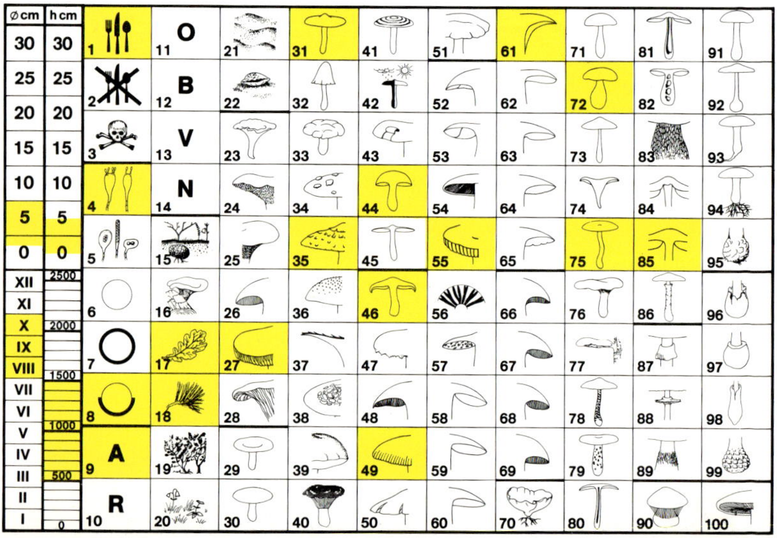

Peppery Milk Cap

Cap convex with margin turning inwards; later expanded and even concave; smooth, white, then tinged with yellow; cracked in dry weather. **Gills** white then yellowish with greenish spots. **Stipe** thick, smooth, white, solid. **Flesh** white. **Smell** faint. **Taste** very sharp. **Milk** white and very sharp. **Spores** white, 7·5–9·5 × 5·5–7 μ; oval. **Habitat** in frondose woods from late summer to autumn; in groups. **Edible/Poisonous** inedible.

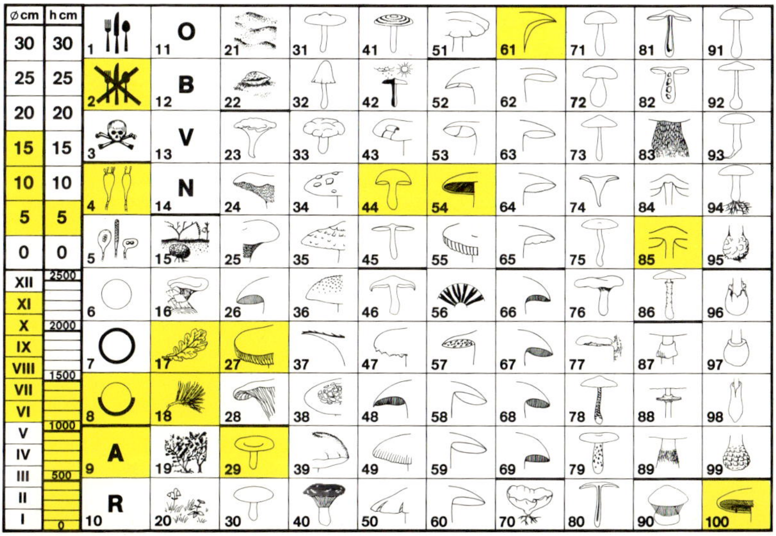

Fleecy Milk Cap or Velvet Cap
Cap concave or even funnel-shaped, often with involute margin; velvety, white then yellowish with ochre spots. **Gills** white then whitish. **Stipe** thick, rather woolly, white then yellowish. **Flesh** firm and white. **Smell** mild. **Taste** sharp. **Milk** in quantity, white then tinged with sulphur yellow. **Spores** white, 9–12 × 7·5–10 μ; globular with warts. **Habitat** in the woods, in groups of several specimens. **Edible/Poisonous** inedible.

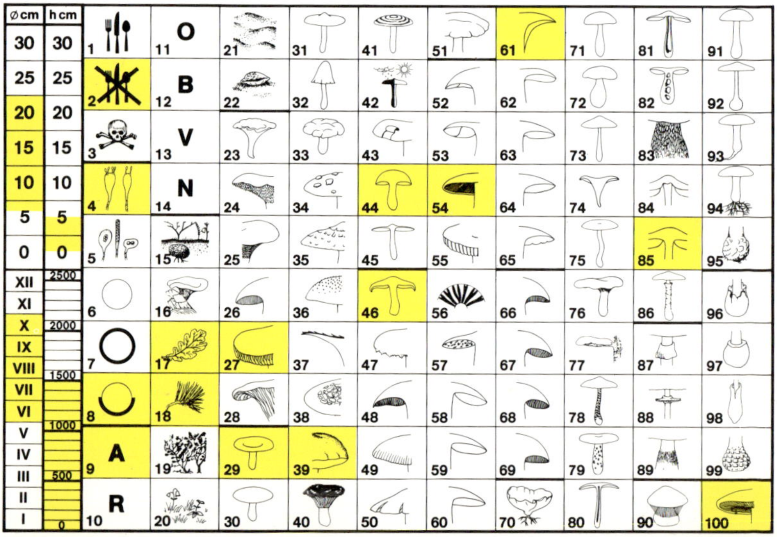

Milk-white Russula

Cap soon becomes depressed or funnel-shaped, margin involute or drooping; matt, fibrillose; white, occasionally tinged with ash or yellow; almost always soiled with earth or vegetal remains. **Gills** white, dewy, with greenish tinge. **Stipe** thick, often like an upturned cone; white, often with brownish spots; matt, full. **Flesh** firm white. **Smell** slightly salty. **Taste** considerably sharp. **Spores** white, 9–12 × 7–9 μ; oval with droplets and warts. **Habitat** in the woods, in groups, quite common. **Edible/Poisonous** to be avoided although edible, as of very poor quality.

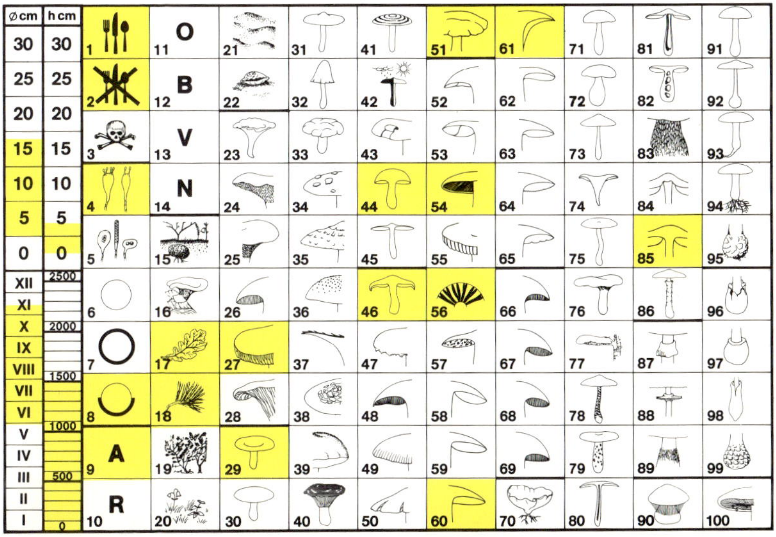

Lead Puffball
Globular without stipe, attached to the soil by root-like filaments; smooth, white then lead-like, cracked in dry weather. **Flesh** firm and white, later granular and brownish, exuding from an opening in the membraneous peridium. **Spores** sooty or greenish, 5·5–6·5 × 4·5–5 μ; ovoid. **Habitat** in open spaces particularly during the rainy seasons. **Edible/Poisonous** inedible when ripe, edible while still very young.

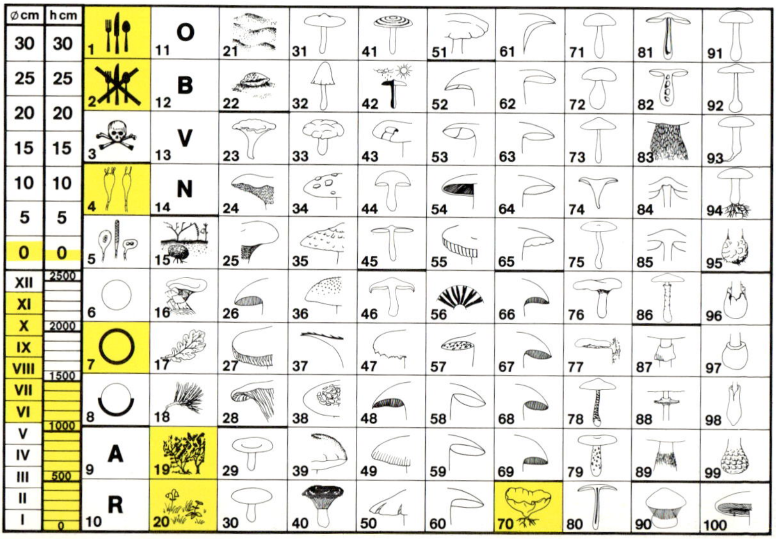

Meadow Puffball
Syn. *Vascellum pratense* (Pers, per Pers.) Kreis
Globular or like an inverted pear, no stipe; covered by a membrane, first white and granular, then smooth and yellowish, and finally open at the top to let out the ripe spores. **Flesh** firm and white, later powdery and greenish-brown. **Smell** mild. **Taste** pleasant. **Spores** brownish-olive, 3–4·5 μ; globular. **Habitat** in the meadows both on low and high ground. **Edible/Poisonous** inedible when ripe, edible while the flesh is still firm and white.

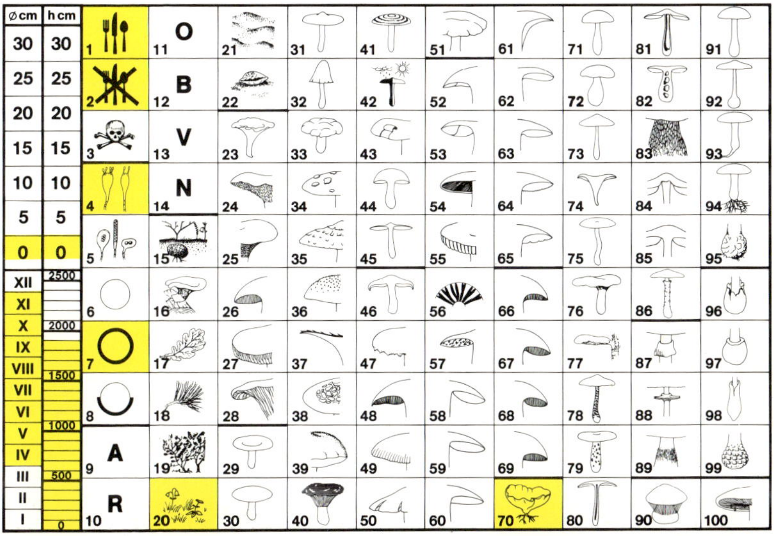

Amanita vaginata Bulliard ex Fries

Grisette
Cap globular or oval, then conic or campanulate, finally flat-umbonate; grey, but variously-coloured in the varieties; occasionally with large pieces of the general veil; margin deeply striate. **Gills** white. **Stipe** slender, white, without ring, with sheathing volva. **Flesh** fragile, white. **Smell** faint. **Taste** faint. **Spores** white, 9–13 μ; globular. **Habitat** usually in the woods, occasionally on the outskirts. **Edible/Poisonous** edible; of good quality after cooking.

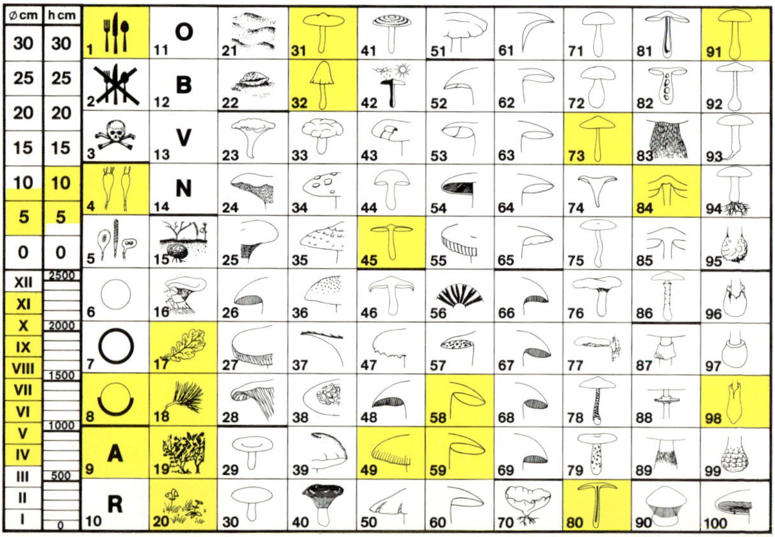

Cap often considerably irregular, as if deformed or compressed here and there by fingertips; with some whitish warts; striate margin; whitish, ash, or brownish. **Gills** white. **Stipe** showing fragments of the partial veil rather than a ring; often curved with the foot several centimetres in the soil; volva soft and variously broken; fragile, soon tubular. **Flesh** white. **Smell** faint. **Taste** mild. **Spores** white, 11–13 × 8·5–10·5 μ; oval with apex. **Habitat** in summer, under oaks, in groups, rare. **Edible/Poisonous** edible; excellent once cooked.

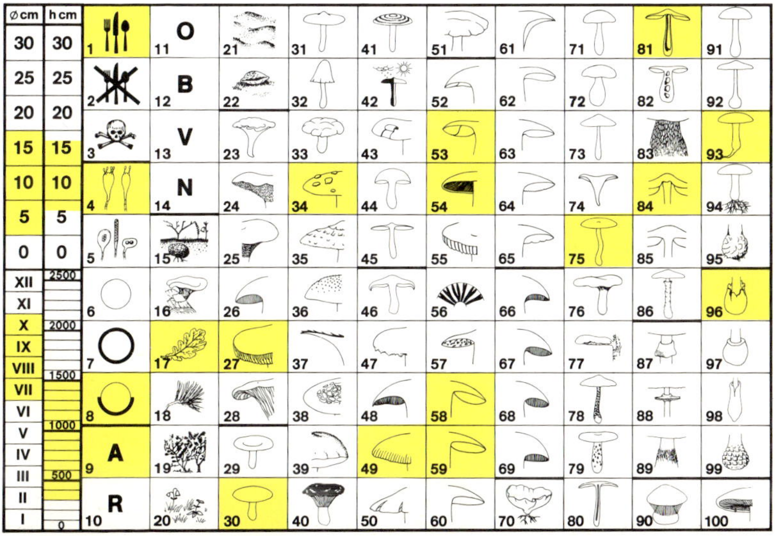

Handsome Volvaria
Cap grey or purple-grey; viscid cuticle which dries up with age and cracks radially. **Gills** pink and then cocoa-brown. **Stipe** tapering upwards, silky, white then ash, solid. **Flesh** white, flaccid in the cap, fibrous in the stipe. **Spores** yellowish-pink, 11–17 × 7·5–9 µ; elliptical-oblong. **Habitat** on soils rich in humus and humidity. **Edible/Poisonous** inedible, although not toxic, as not very good.

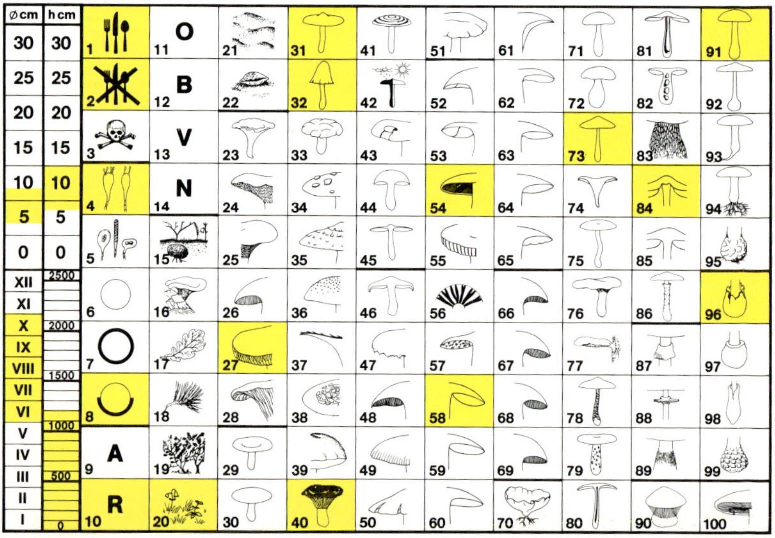

Fawn-coloured Pluteus
Syn. *Pluteus atricapillus* (Secr.) Sing.
Cap sooty-brown, fibrillose, viscid in wet weather. **Gills** white, then pink and later tinged with rust. **Stipe** whitish with brownish fibrils. **Flesh** considerably flaccid, white. **Smell** of turnip. **Taste** of turnip. **Spores** rusty-pink, 7–8 × 4·5–6 μ; oval. **Habitat** on the bole of various trees, on rotting branches and corn stalks. **Edible/Poisonous** edible, but of poor quality.

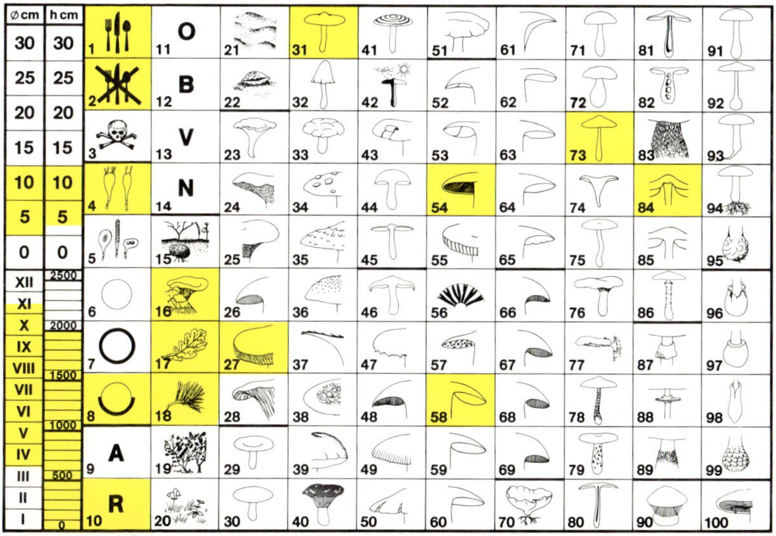

Common Ink Cap
Cap oval then campanulate; ash sometimes tinged with ochre or somewhat sooty; now smooth, now fibrillose and scaly at the top. **Gills** whitish, then brownish and finally dissolving into black liquid. **Stipe** white, shiny; foot almost bulbous and marginate, sometimes with root-like appendix. **Flesh** whitish then blackish. **Smell** pleasant. **Taste** appealing. **Spores** sooty-black, 7–11 × 6–6·5 μ; elliptical. **Habitat** in early spring and up to late autumn, on soil enriched with organic substances, in tufts of several specimens. **Edible/Poisonous** edible with great care; should be cooked, and eaten without alcohol.

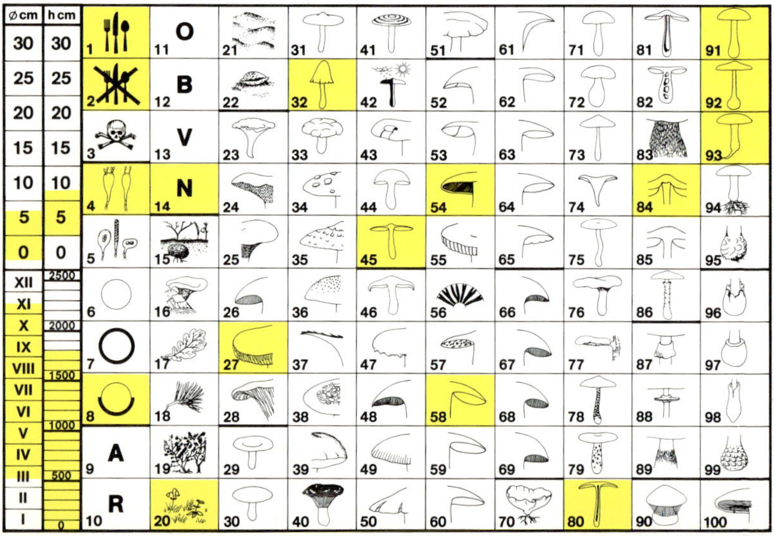

Cap shiny, fibrillose, sometimes cracked; white tinged with ash or yellow. **Gills** yellowish tinged with olive, with a paler edge. **Stipe** white, fribrillose, silky, occasionally tinged with yellow. **Flesh** fragile, white. **Smell** earthy. **Taste** earthy. **Spores** fawnish, 10–14 × 5·5–7·5 μ; elliptical or kidney-shaped. **Habitat** in the woods, from June until November. **Edible/Poisonous** poisonous.

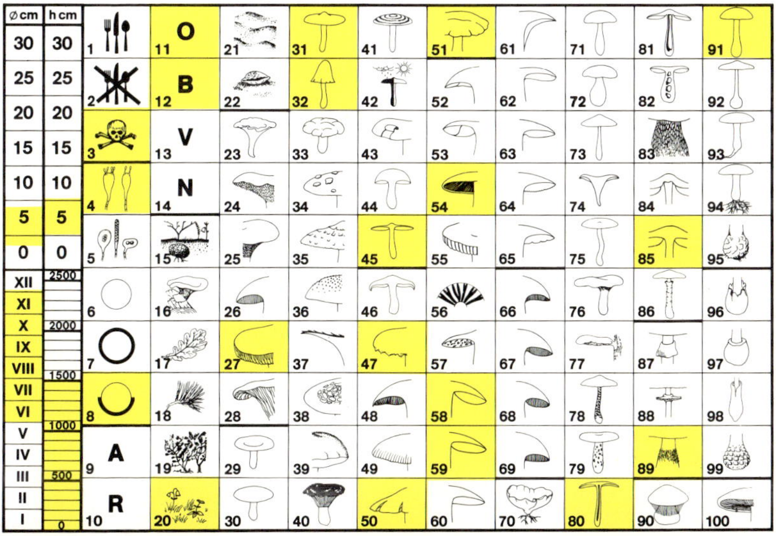

Nolanea pascua Persoon

Syn. *Nolanea staurospora* Bresadola; *Rhodophyllus staurosporus* (Bres.) J. Lange
Cap campanulate, but also plane while still umbonate; margin turning downwards, smooth; from nut-brown to brown; soft, fibrillose cuticle. **Gills** ash, then pink and finally rusty. **Stipe** woolly, concolorous with the cap, the foot covered in cotton-wool-like substance; fibrous then fistulous; marked by a deep furrow from top to bottom. **Flesh** grey. **Smell** floury and faint. **Taste** floury and mild. **Spores** pink, 9–11 × 7–8·5 μ; polygonal. **Habitat** in rainy weather, from spring to autumn, in woods, meadows, in the mountains or on plains, in small groups. **Edible/Poisonous** poisonous.

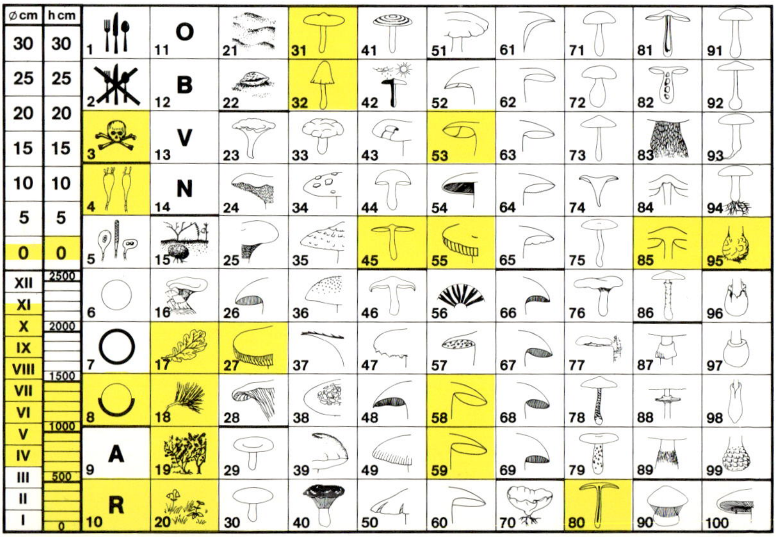

Syn. *Tricholoma pardinum* Quél.
Cap almost globular or campanulate, involute margin; then expanded with protruding top; scaly, ash sometimes tinged with sooty-brown. **Gills** white, tinged with cream; later with yellowish-green iridescence. **Stipe** with occasionally swollen foot; whitish, fibrillose. **Flesh** firm in the cap, fibrous in the stipe; whitish. **Smell** floury. **Taste** floury. **Spores** white, 8–10 × 6–7 μ; elliptical. **Habitat** in the woods, in groups. **Edible/Poisonous** poisonous.

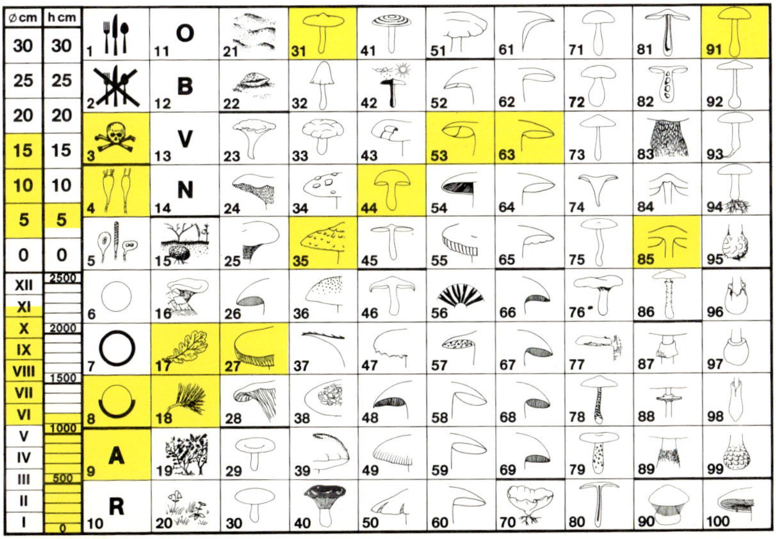

Earth Tricholoma

Cap generally umbonate, rarely depressed; grey, sometimes tinged with brown, with lighter margin; completely covered by darker scales. **Gills** whitish then ash-grey. **Stipe** rarely with scales, usually smooth; white then ash-grey; first full, then with spongy marrow. **Flesh** fragile, whitish or ash-grey. **Smell** mild. **Taste** mild but bitter in old specimens. **Spores** white, 5–8 × 3·5–5 μ; elliptical. **Habitat** from July to December under conifers, in groups or tufts. **Edible/Poisonous** edible; of good quality when very young.

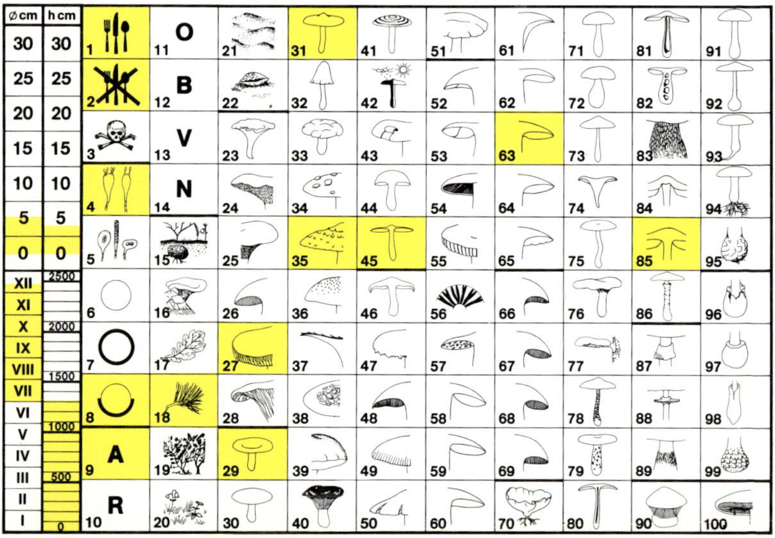

Cap fleshy, slightly depressed towards the end, with unrolled margin; fibrillose or scaly; ash-grey to brown. **Gills** adnate to hooked, rather massive but not very thick; white then grey. **Flesh** firm, ashy-white, red in the foot of the stipe. **Smell** of flour. **Taste** of flour. **Spores** white, 6·2–8·5 × 4·6–5·8 μ; elliptical-oval, smooth. **Habitat** in summer and autumn, under frondose trees. **Edible/Poisonous** edible.

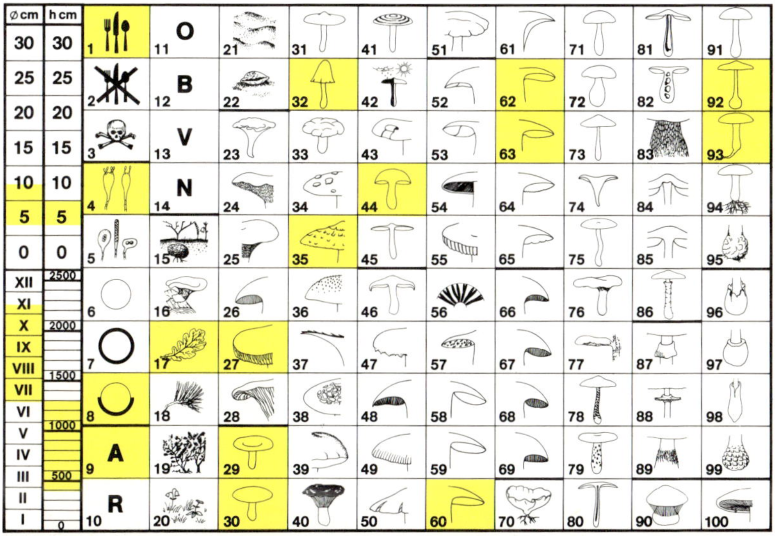

34 Lyophyllum aggregatum (Schaeffer ex Fries) Kühner

Fried-chicken Mushroom
Syn. *Lyophyllum decastes* (Fr.) Sing.
Cap smooth or woolly, from ash-grey to brownish; radial fibrils of darker colour. **Gills** white then creamy. **Stipe** often irregular, the foot linked with that of the other specimens in the same tuft; white, smooth, shiny; full. **Flesh** firm in the cap, fibrous and elastic in the stipe; white. **Smell** pleasant. **Taste** slightly bitter. **Spores** white, 5–7 μ; globular. **Habitat** in the autumn, sometimes in spring; around tree boles or on underground roots, in tufts, sometimes huge. **Edible/Poisonous** edible; of good quality.

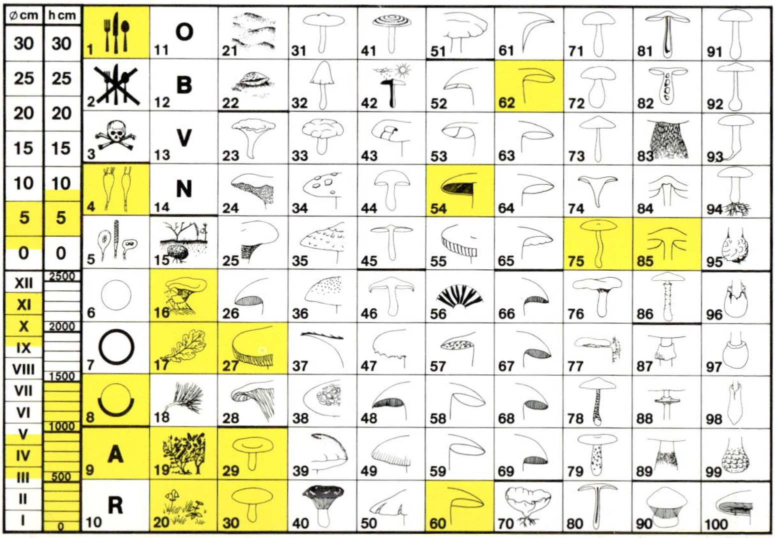

Clouded Clitocybe

Cap occasionally slightly convex in the centre, usually plane and finally convex; smooth, shiny, often covered in white down in the centre; ash or brownish, but sometimes white. **Gills** creamy. **Stipe** often attenuating towards the cap, the foot swollen and covered in cotton-wool; fibrillose, concolorous with the cap or slightly paler; full, later hollow. **Flesh** somewhat flaccid in the cap, thick in the stipe; white. **Smell** sharp and unpleasant. **Taste** acrid-bitter. **Spores** white, 5–9 × 3–4·5 μ; elliptical. **Habitat** in the woods in autumn, in groups; very common. **Edible/Poisonous** slightly poisonous if raw; edible, although not good when cooked.

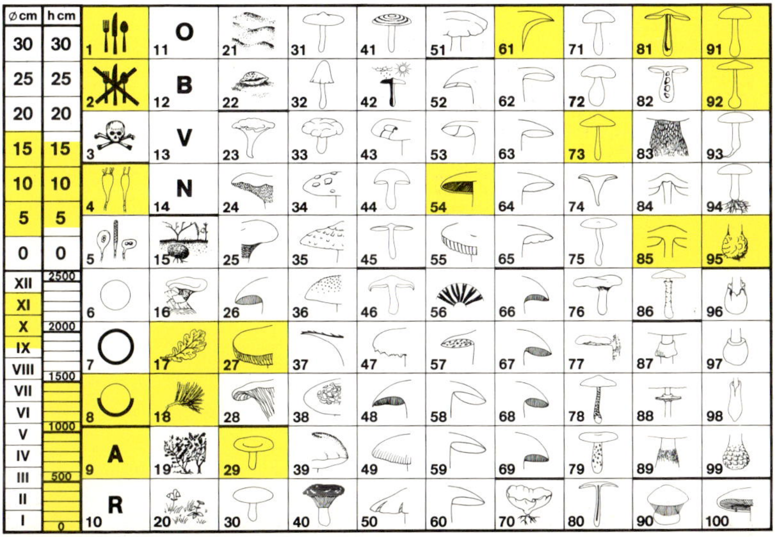

Clitocybe odora Bulliard ex Fries

Aniseed Clitocybe

Cap with involute margin, later expanded; cuticle smooth, green/greenish/bluish. **Gills** greenish or tinged with yellow. **Stipe** a little paler than the cap, the foot covered in white down and surrounded by mycelium cords. **Flesh** greenish. **Smell** of aniseed. **Taste** mild. **Spores** white with pink tinge, 6–7·5 × 3–4 μ; elliptical. **Habitat** in small groups in woods, almost up to winter. **Edible/Poisonous** edible with caution; should always be mixed with other mushrooms as a condiment.

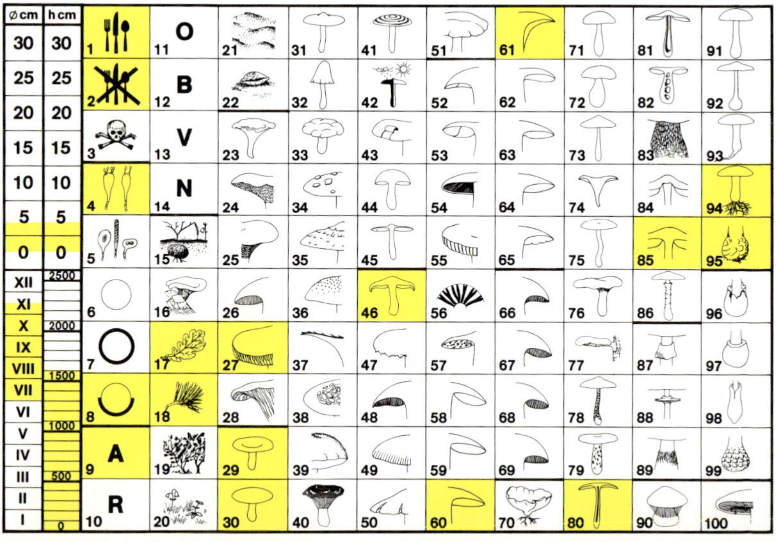

Sweetbread Mushroom or The Miller

Cap convex, with involute and lobate margin; then expanded and often concave; dry and soft, sometimes downy; whitish, sometimes shaded with ash or cream. **Gills** white then pink. **Stipe** narrowing downwards to a pointed foot; sometimes curved and eccentric; white; downy near the cap; full. **Flesh** tender and fragile; white. **Smell** of flour. **Taste** pleasant. **Spores** pink, 10·5–14 × 5–6 μ; elliptical-oblong. **Habitat** from spring to autumn on any soil, in groups or small tufts. **Edible/Poisonous** edible; of good quality; can be preserved in various ways.

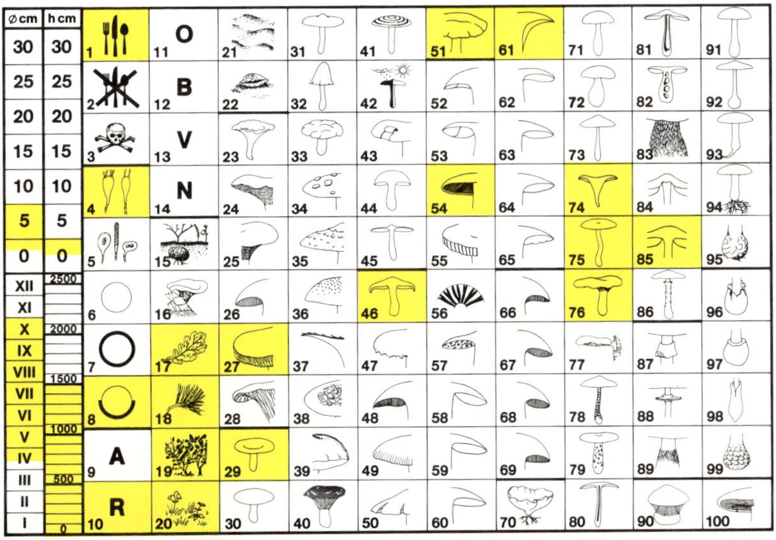

Oyster Mushroom

Cap semicircular, plane, slightly concave; margin first involute then plane; smooth and slightly downy; grey or brownish or sooty. **Gills** white then creamy. **Stipe** eccentric or lateral with a thin foot; white; grey, downy and velvety near the cap. **Flesh** firm in the cap, fibrous in the stipe; white. **Smell** slight, of flour. **Taste** good. **Spores** ash-pink, 8–11 × 3–4·5 μ; cylindrical-oblong. **Habitat** from March to September on the trunks of frondose trees, rare on conifers; in tufts. **Edible/Poisonous** edible; of good quality while still young.

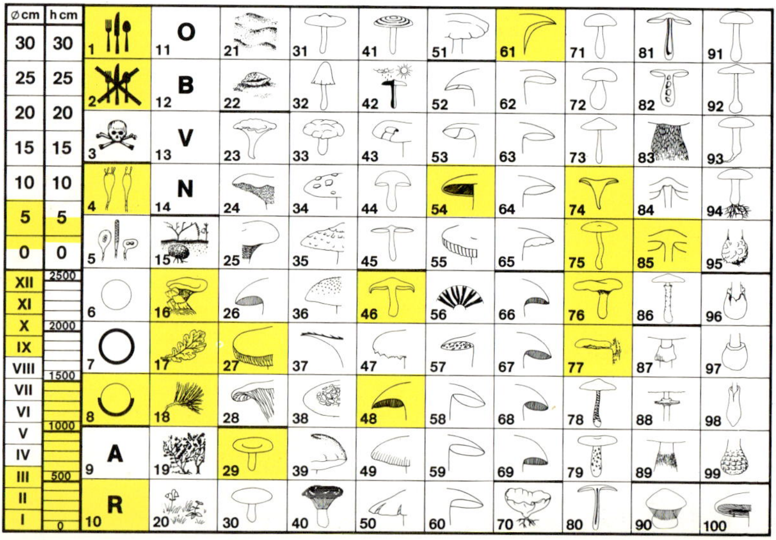

Cap viscid in wet weather, otherwise dry; difficult to peel; whitish, then spotted with blackish, finally black. **Gills** occasionally forked; white, then yellowish spotted with black and finally black. **Stipe** concolorous with the cap or slightly lighter; full, then with marrow. **Flesh** hard, white becoming black quickly when exposed to the air. **Smell** faint. **Taste** bitter. **Spores** white, 7–9 × 6–6·5 μ; globular, reticulated, with warts. **Habitat** in mixed woodlands. **Edible/Poisonous** inedible; of very poor quality.

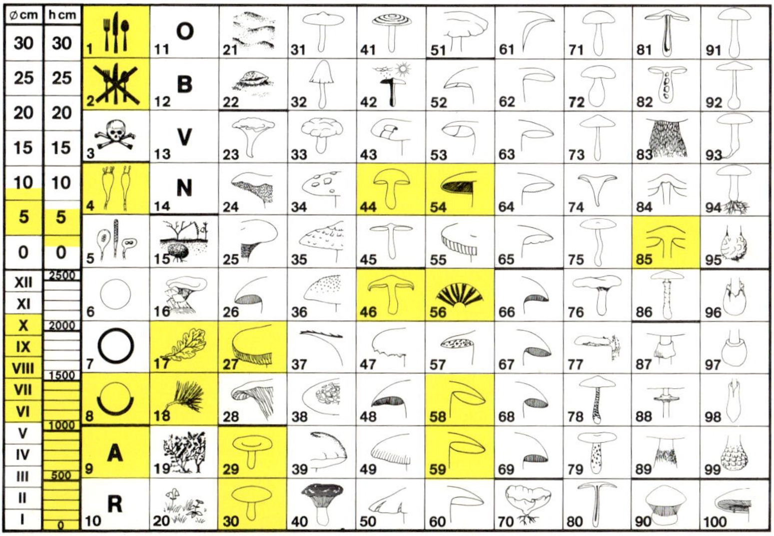

Hygrophorus marzuolus (Micheli ex Fries) Bresadola

Spring Wax Cap

Cap convex, then irregularly expanded and often even depressed; smooth, whitish, or ash/grey/sooty, sometimes tinged with ochre. **Gills** waxy, white and finally ash. **Stipe** usually thick, often strangely bent; shiny, downy; white or ash-grey; full. **Flesh** firm, white, tinged with ash-grey at the margin. **Smell** very faint. **Taste** faint. **Spores** white, 6·5–8 × 5–6·5 μ; elliptical. **Habitat** at the end of winter and in spring, in woods, heathlands, in groups; rare, but abundant in its colonies. **Edible/Poisonous** edible; of good quality.

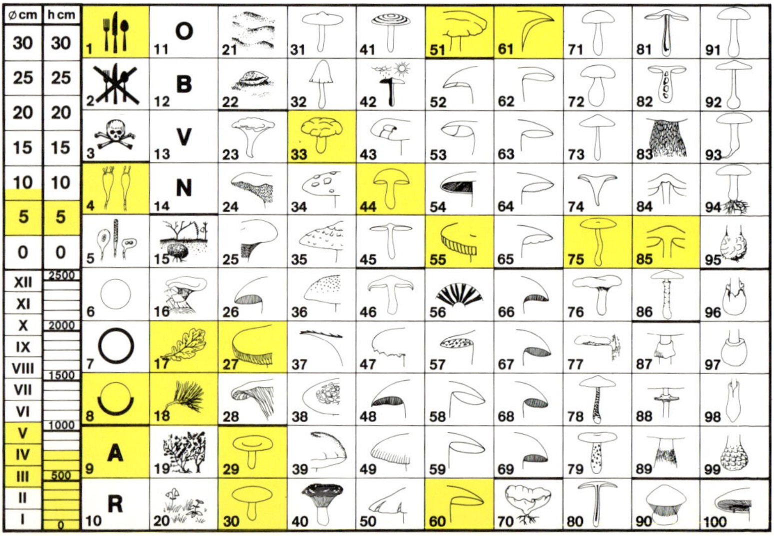

Syn. *Gyroporus cyanescens* (Bull. ex Fr.) Quélet

Cap whitish to pale ochre, sometimes tinged with olive; flaky to scaly. **Tubes** 1·5–2·5cm long, white, becoming blue in contact with air, then olive brown. **Pores** round, yellowish-white, blue when pressed. **Stipe** slightly swollen, concolorous with the cap but white at the top; fragile, finally hollow. **Flesh** firm in the cap, marrow-like in the stipe; white, blue when cut, then yellowish and finally sooty. **Smell** pleasant. **Taste** pleasant. **Spores** yellowish, 8–13 × 4–8 μ; elliptical. **Habitat** on sandy soils under broadleaf trees, in small groups or singly. **Edible/Poisonous** edible; of good quality.

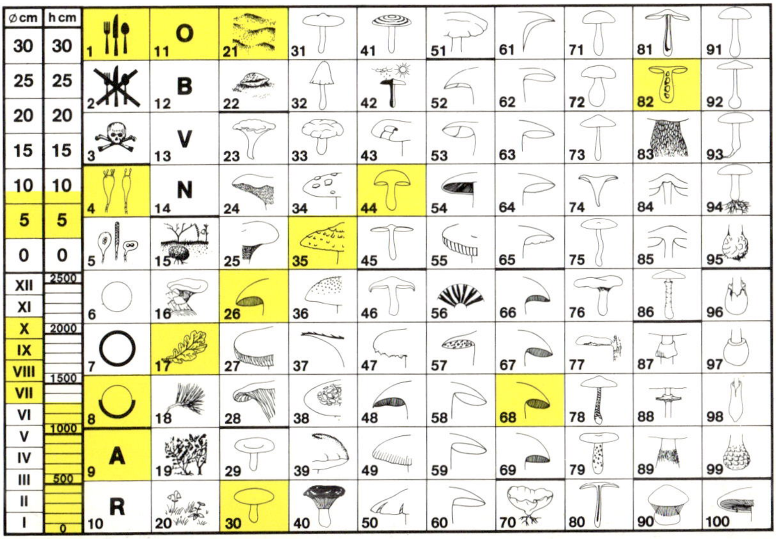

Devil's Bolete

Cap with wavy margin; velvety, then smooth, dry and cracked; from whitish to ash-grey with olive tinge. **Tubes** 2–3cm long, at a tangent with the stipe, then free; yellow then olive. **Pores** small and round, yellow then red and orange, paling with age; when touched they become blue then glaucous. **Stipe** oval, later still swollen at least at the foot; red, but brownish at the foot; with a red, pink, brown or olive reticulation, sometimes replaced by spots of the same colours; becoming blue then blackish when touched; often hollow in the foot. **Flesh** soon spongy, whitish with yellow spots; exposed to the air it turns bluish then whitish again. **Smell** pleasant. **Taste** bitter after some time. **Spores** olive-brown, 11–15 × 5–6 μ; fusoid. **Habitat** in sunny glades. **Edible/Poisonous** poisonous.

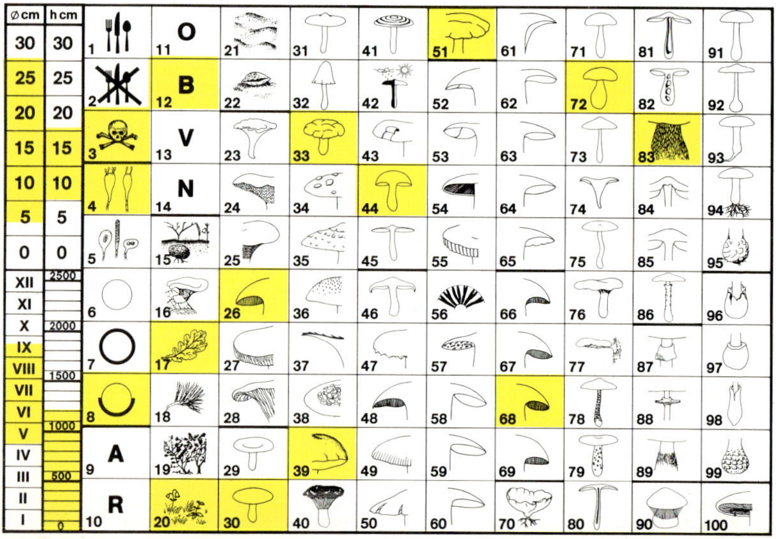

Poplar Bolete

Syn. *Leccinum duriusculum* (Kalchbr. & Schulz. apud Fr.) Sing.

Cap with cuticle growing beyond the margin; velvety, ash-grey, sometimes grey or smoky-ochre with olive tinge; paling in dry weather. **Tubes** 1·5–2·5cm long, slightly tangent or free; whitish. **Pores** tiny, round, whitish then ash-grey. **Stipe** claviform or fusoid; whitish, occasionally with glaucous shading and vertical wrinkles; covered in smoky scales. **Flesh** very firm in the cap, hard and fibrous in the stipe; when broken remains white while damp then pink and finally sooty. **Smell** mild. **Taste** sweetish. **Spores** brownish, 12–20 × 5·6 μ; fusoid, slightly curved. **Habitat** under poplars and birches. **Edible/Poisonous** edible; of good quality.

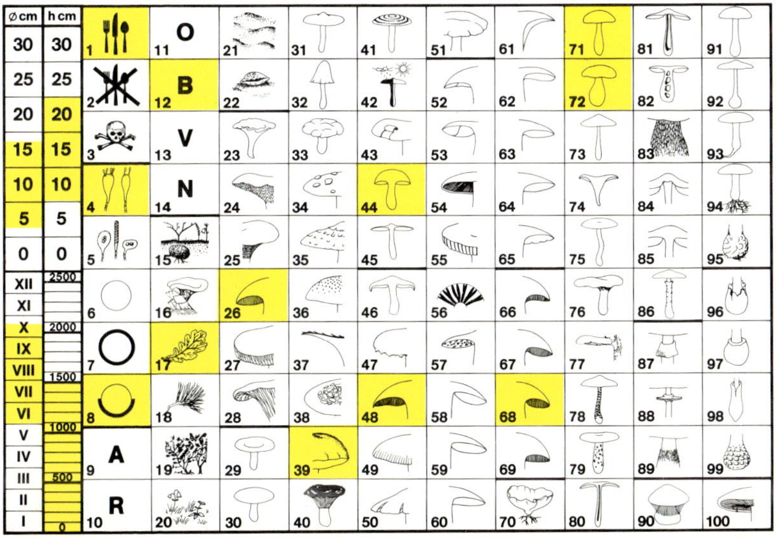

Polyporus confluens Albertini et Schweinitz ex Fries

Caps numerous, variously shaped, with wavy surface and lobed margin; smooth, yellowish or tinged with pink, sometimes ochre; inseparable and later fragmented cuticle. **Tubes** short, 0·1–0·3cm, decurrent almost to the foot; white. **Pores** small, round, white then yellowish. **Stipes** thick, irregular, widening towards the cap, eccentric; joining at the base in one whole mass; white then yellowish; full. **Flesh** firm, white. **Smell** faint. **Taste** acrid-bitter, bitter in old specimens. **Spores** white, 4–5·5 × 3–4 μ; elliptical-globular. **Habitat** in the grass, not far from conifers. **Edible/ Poisonous** edible, but only if thoroughly immature, deprived of the cuticle with a sharp knife and preboiled.

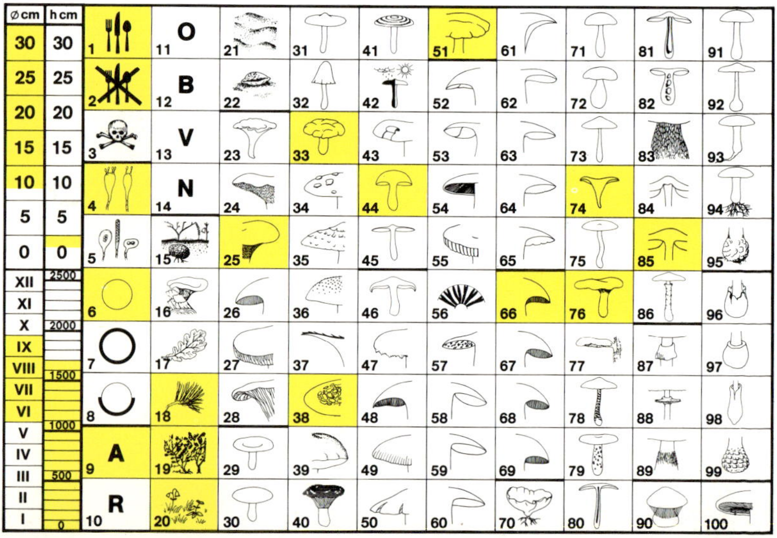

Polyporus umbellatus Persoon ex Fries

A thick, whitish, often wrinkled trunk supports almost cylindrical branches which in turn are subdivided into smaller ones; the latter bear the **caps**, often more than one hundred, convex and umbonate, later depressed (2–4cm in diameter), ochraceous, scaly, often cracked, with wavy and often split margin. **Tubes** short, decurrent, white. **Pores** tiny, roundish, white. **Flesh** tender and slightly fibrous; white. **Smell** of flour. **Taste** sweetish. **Spores** white, 7–11 × 3–4 μ; elliptical. **Habitat** on stumps of frondose trees, in tufts. **Edible/Poisonous** edible; of good quality; can be preserved in oil or vinegar.

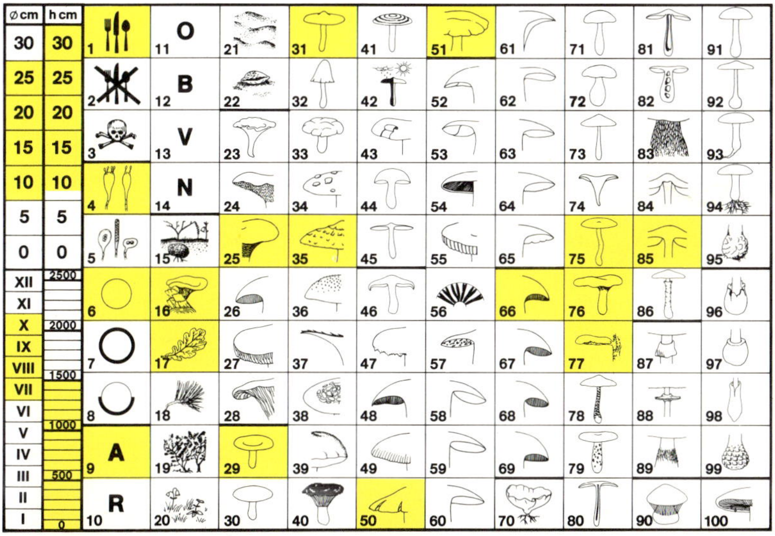

Round Morel

Cap globular or ovoid, hollow, honeycombed surface, ochre with lighter ribs. **Stipe** hollow, cylindrical, irregular, often with swollen foot; deeply furrowed, floccose, whitish. **Flesh** fragile, waxy, white. **Smell** faint. **Taste** pleasant. **Spores** tinged with ochre, 20–24 × 12–16 μ; elliptical. **Habitat** on sandy soils in spring. **Edible/Poisonous** edible; of good quality.

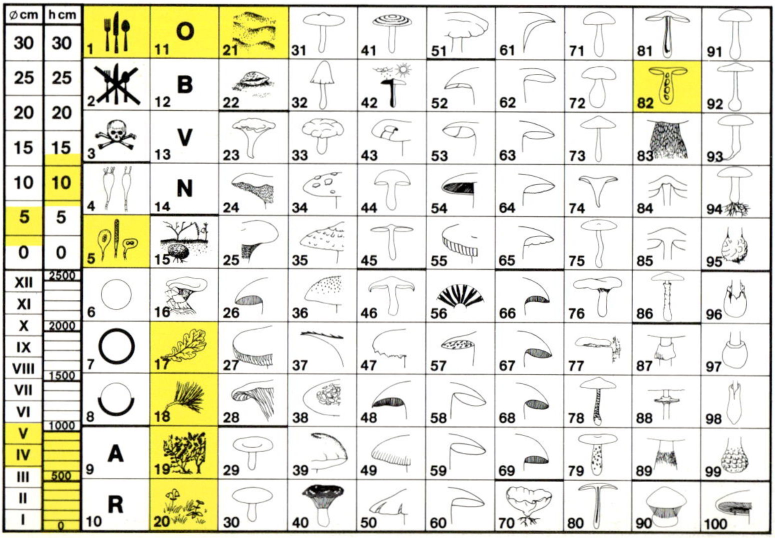

Cap hollow, conical or oval; honeycombed, yellowish, ochre or brownish; almost continuous ribs descending from the cap, lighter in colour. **Stipe** irregular, rough, white or whitish; hollow. **Flesh** fragile. **Smell** pleasant. **Taste** faint. **Spores** ochraceous, 18–22 × 12–16 μ; elliptical. **Habitat** either in woods or in the open in spring. **Edible/Poisonous** edible; of good quality.

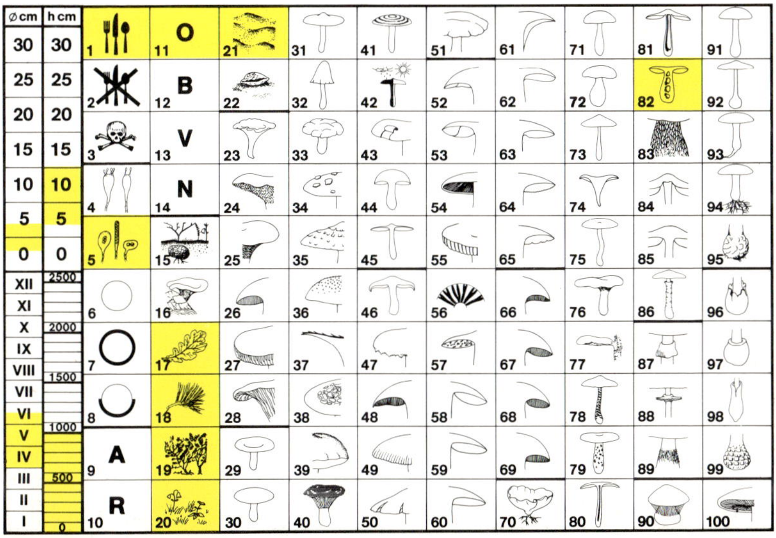

Syn. *Pluteus aurantiorugosus* (Trog.) Sacc.
Cap with overlapping margin, often striate; red in wet weather, pale orange in dry weather; usually umbonate; radially fibrillose, occasionally grooved. **Gills** free, sometimes wavy; whitish then tinged with pinkish-yellow with lighter edge. **Stipe** often swollen at the foot; slightly bent, rarely straight; rough, white then yellowish, particularly towards the foot; full, then fistulous. **Flesh** fragile; yellowish, but whitish in the centre of the stipe. **Smell** faint. **Taste** sweetish. **Spores** rosy, 5·4–6·6 × 4·2–5 μ; globular. **Habitat** in summer and autumn on the stumps of frondose trees, or on wood shavings, in tufts. **Edible/Poisonous** edible in small quantities; the effect of large quantities has not been tested yet.

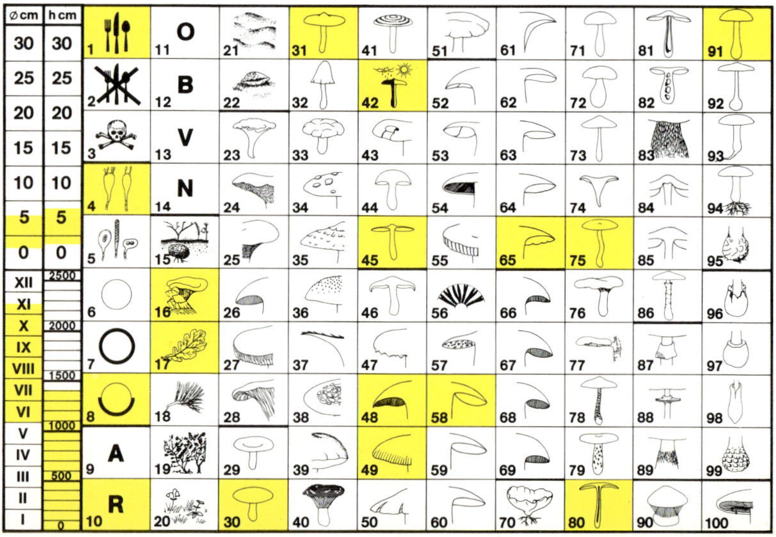

Brick-red Hypholoma

Syn. *Hypholoma sublateritium* (Fr.) Quélet

Cap hemispherical, then plane, but generally slightly convex; orange, reddish at the top, yellowish at the margin. **Gills** greyish-yellowish, then greyish-olivaceous. **Stipe** sometimes curved, concolorous with the cap, with pseudo-ring left by the cortina. **Flesh** yellowish. **Smell** mild. **Taste** bitter. **Spores** olive-brownish, 6–7·5 × 3–4·5 μ; elliptical to oval. **Habitat** from spring to late autumn on live or dead wood; in tufts. **Edible/Poisonous** poisonous; causes serious intestinal troubles.

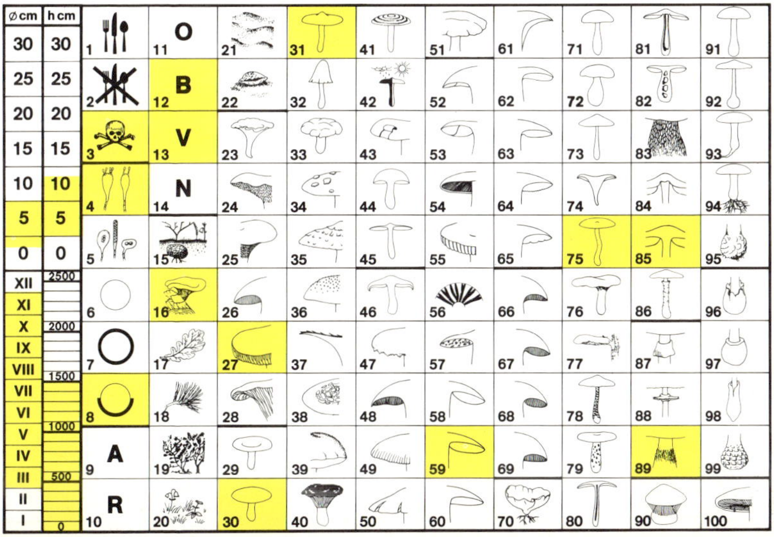

Red-staining Inocybe

Cap whitish, then yellowish, and finally reddish or freckled; radially fibrillose, dry, silky, often cracked half-way up the radium. **Gills** white or rosy, then rusty or brownish-olivaceous, with irregular, flaky and lighter edge. **Stipe** white or whitish, spotted with reddish; fibrillose but floccose or granular near the cap where it occasionally shows small fragments of the cortina. **Flesh** white. **Smell** faint, of fruit. **Taste** mild. **Spores** rust, 10–15 × 6–8 μ; elliptical or kidney-shaped. **Habitat** especially under lime trees, in groups. **Edible/Poisonous** poisonous.

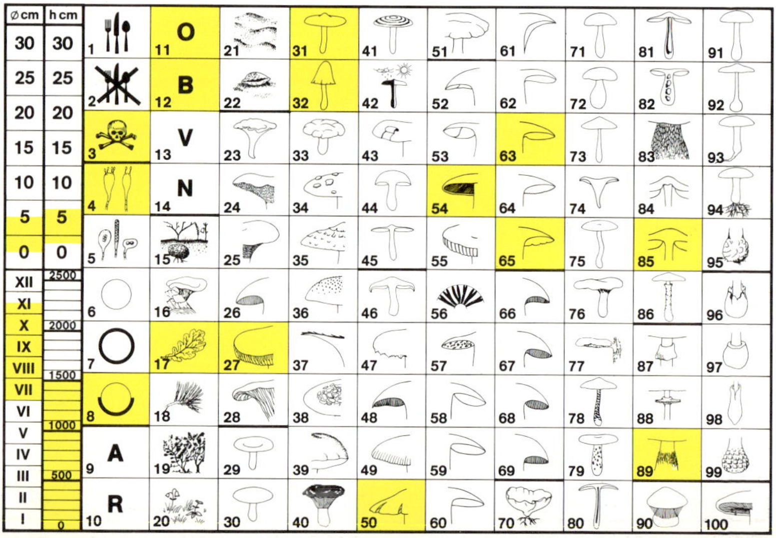

Woolly Milk Cap or Griping Toadstool

Cap with involute margin, fringed even when assuming a cup-like or funnel-like shape; woolly, yellowish-pink or pinkish-ochraceous, with paler concentrical areas. **Gills** creamy. **Stipe** occasionally slightly scobiform with floccose area near the gills; concolorous with, or paler than, the cap. **Flesh** whitish. **Smell** faint. **Taste** sharp. **Milk** white and sharp. **Spores** white flushed with cream or pink, 5–10 × 4·5–6·5 μ; elliptical, with warts. **Habitat** under birch trees, rare under other broad-leafed trees, in groups. **[redacted]/Poisonous [redacted]e.**

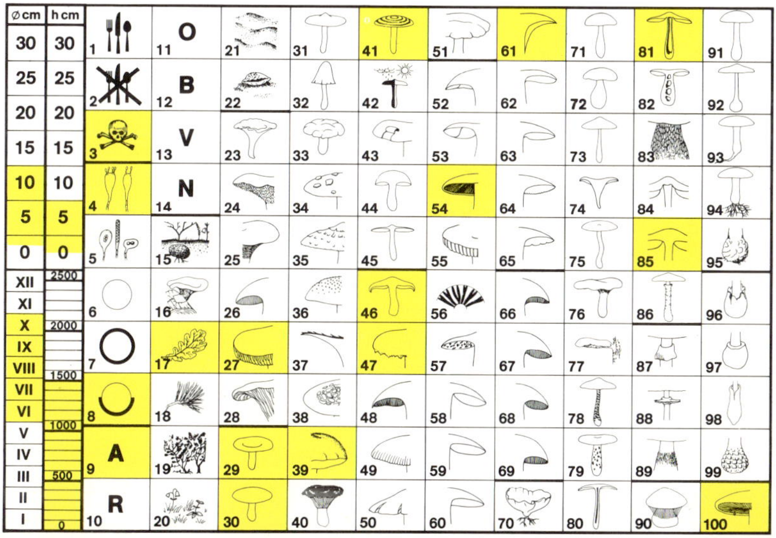

Saffron Milk Cap
Cap quickly depressed, maintaining involute margin; viscid in wet weather; fibrillose and almost velvety; orange, sometimes with concentric zones tinged with greenish-blue. **Gills** creamy then orange, with greenish spots. **Stipe** often pitted, concolorous with cap. **Flesh** fragile, orange, slowly turning greenish when exposed to the air. **Smell** of fruit. **Taste** sweetish. **Milk** orange, copious and sweetish. **Spores** creamy, 7–9 × 6–7 μ; globular, reticulated. **Habitat** under conifers, in large groups. **Edible/Poisonous** edible; of good quality.

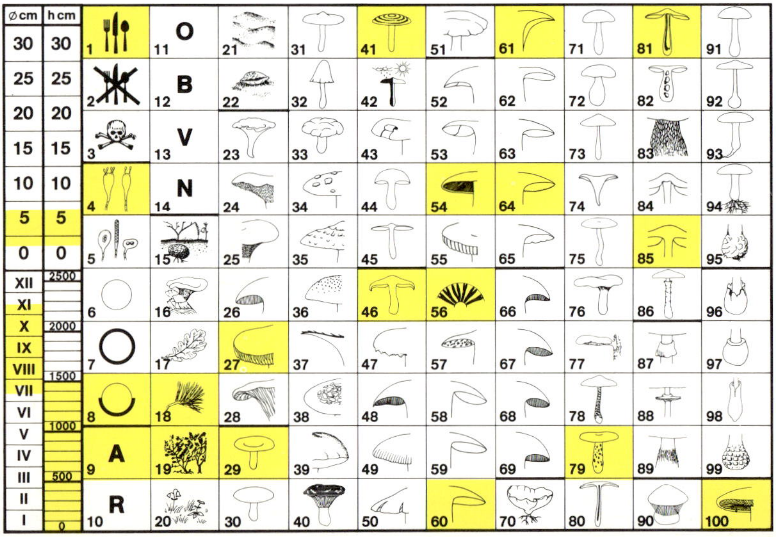

Crimson Wax Cap

Cap initially with involute margin, later often with up-turned margin, fissured; orange-red or vermilion-red, sometimes paler on the margin. **Gills** distant, thick and broad; yellowish-pink or reddish with yellowish edge. **Stipe** often with pointed foot; concolorous with cap or slightly paler; fibrillose; at times with one or two deep furrows; fistulous. **Flesh** fragile, yellow. **Smell** faint. **Taste** pleasant. **Spores** white, 8–12 × 4–6 μ; elliptical, smooth. **Habitat** in late summer, on hill and mountain pastures, in groups or singly. **Edible/Poisonous** edible.

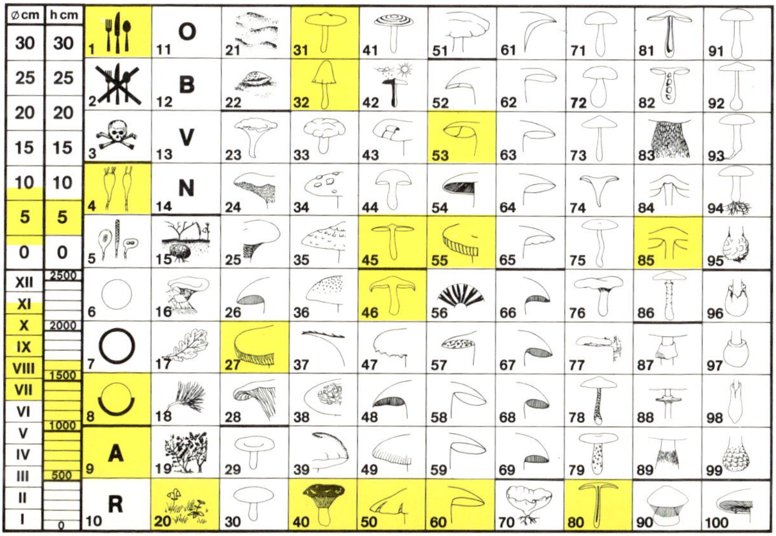

Syn. *Leccinum aurantiacum* (Bull. ex St. Am.) S. F. Gray
Cap hemispherical, then slightly convex, with overlapping cuticle; velvety; orange. **Tubes** 2–4cm long, adnexed to free; ash-grey. **Pores** small, roundish, white then ash-grey and ferruginous. **Stipe** tall, slightly swollen, often marked by a deep furrow; ash-grey with orange scales; turning greenish when bruised; full. **Flesh** soon becomes soft in the cap, fibrous in the stipe; white, turning pink then blackish in the air. **Smell** mild. **Taste** pleasant. **Spores** ochraceous, 12–19 × 4–4·5 μ; fusoid. **Habitat** under frondose trees, particularly poplars and birches, in groups. **Edible/Poisonous** edible; of good quality.

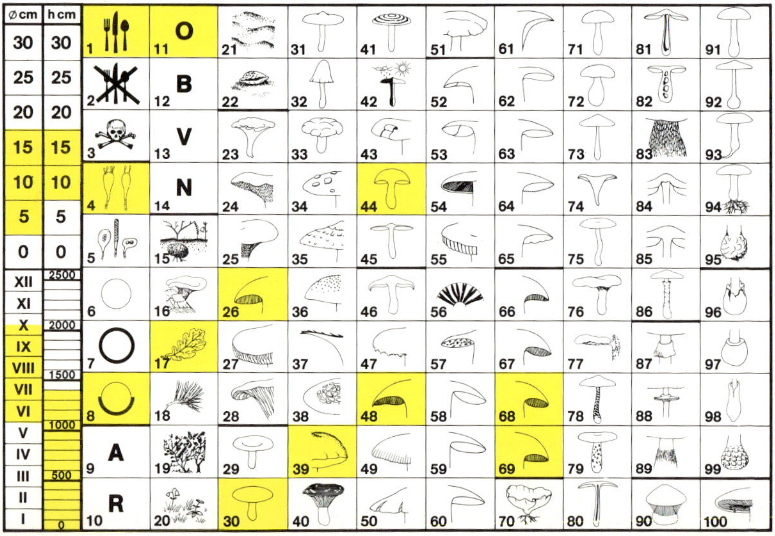

Larch Bolete
Syn. *Boletus grevillei* Klotzsch or *Suillus grevillei* (KL.) Singer
Cap often with a slight umbo even when plane, the margin long involute and fringed; viscid, yellow or orange. **Tubes** 1–2cm long, adnate or slightly decurrent; yellow then rust. **Stipe** fibrillose, yellow, with whitish ring which soon erodes; full. **Flesh** somewhat soft in the cap, fibrous in the stipe; yellowish. **Smell** pleasant. **Taste** pleasant. **Spores** yellowish-olivaceous, 8–11 × 3–4 μ; elliptical. **Habitat** under larches, in groups. **Edible/Poisonous** edible; can be preserved in various ways.

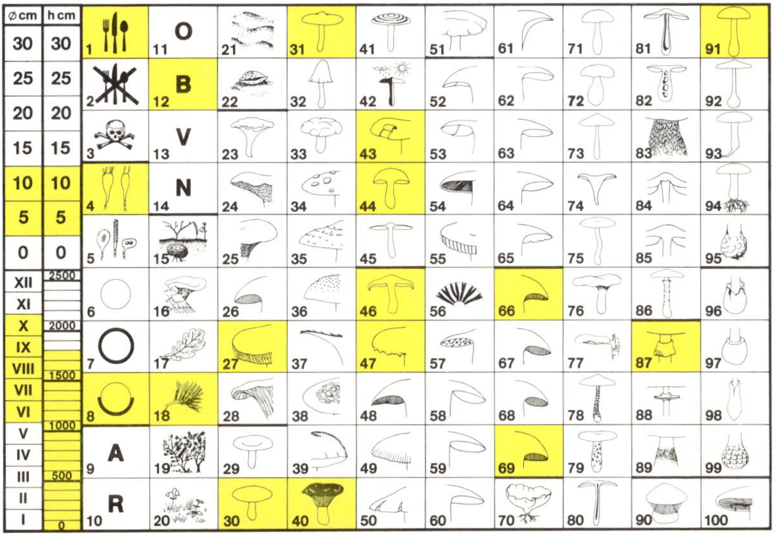

Clavaria subbotrytis (Coker) Corner

Syn. *Ramaria subbotrytis* (Cok.) Corner
A fleshy, yellowish-rose carpophore supports a few vertical branches, orangey-pink and robust, ending in 3–4 small crests of even brighter colour. **Flesh** white, flushed with rose. **Smell** mild. **Taste** slightly bitter. **Spores** yellowish, 7–9·5 × 3–4 μ; elliptical. **Habitat** in the woods, not common. **Edible/Poisonous** edibility unknown.

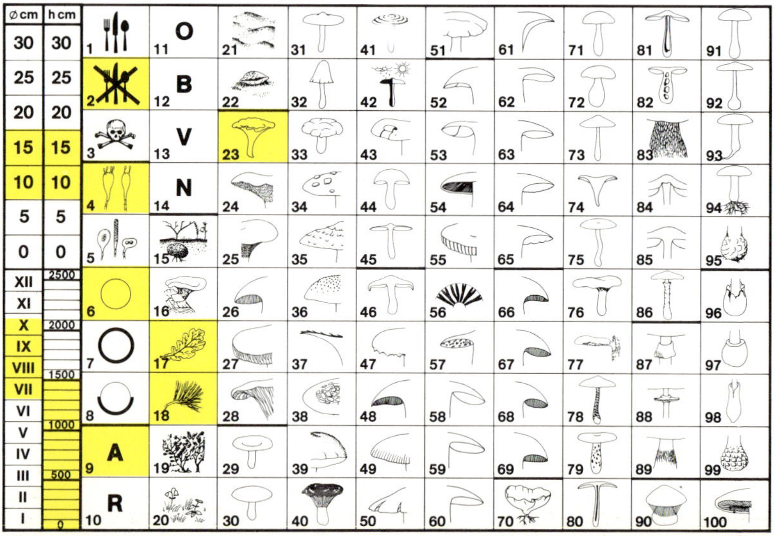

Syn. *Clathrus ruber* Mich. per Pers.
Wrapped at first within a thick volva resembling an egg (2·5–6cm diameter), it emerges in the shape of a spherical net (8–12cm diameter), wide-meshed, polygonal, red, rough, porous and fragile; the internal surface is coated with blackish-olivaceous granules containing the spores; finally becoming mucilaginous and foul-smelling. **Spores** greenish, 5–6 × 3–4 μ; elliptical-oblong. **Habitat** on damp soils. **Edible/Poisonous** edible.

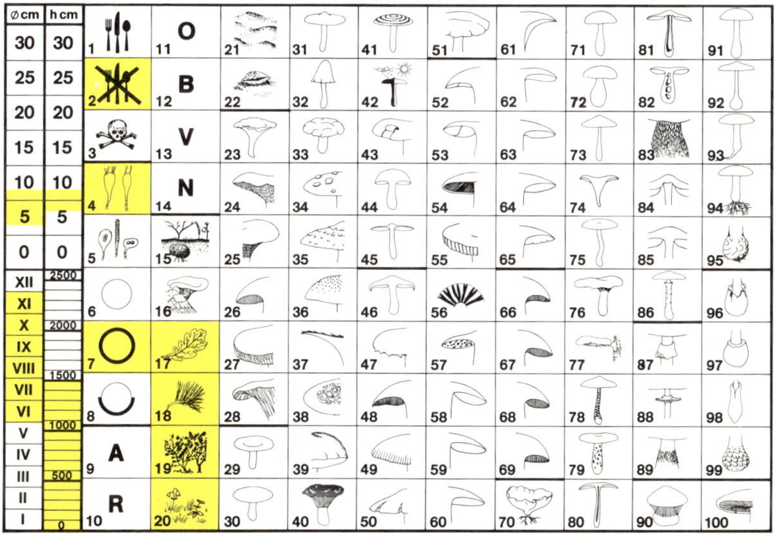

Orange-peel Fungus
Syn. *Aleuria aurantia* (Pers. ex Hook.) Fuckel
Shaped like a small bowl, sometimes supported by a short stipe; wavy or lobed margin; internally smooth, dry, orangey-red; externally lighter, occasionally almost whitish, floccose and downy. **Flesh** thin, orange. **Smell** slightly unpleasant. **Taste** mild. **Spores** white, 16–18 × 9–12 μ; elliptical-fusoid, reticulated. **Habitat** on sandy soils. **Edible/Poisonous** edible, even when raw.

Death Cap

Cap hemispherical, then expanded, greenish to olivaceous, yellowish, ash-grey or whitish; intersected by innate, radial fibrils, darker in colour. **Gills** white then tinged with green. **Stipe** with white fugacious ring; slightly tinged below the ring; bulbous foot, ample volva, membraneous. **Flesh** white and tender. **Smell** imperceptible, but carrion-like in old specimens. **Taste** acid (do not swallow!). **Spores** white, 7–9 × 8–11·5 μ; globular or tear-shaped. **Habitat** anywhere, except in the cold season; extremely common. **Edible/Poisonous** deadly, whether raw or cooked, dried or otherwise treated; lethal dosage about one gram of fresh flesh for each kilo of adult weight.

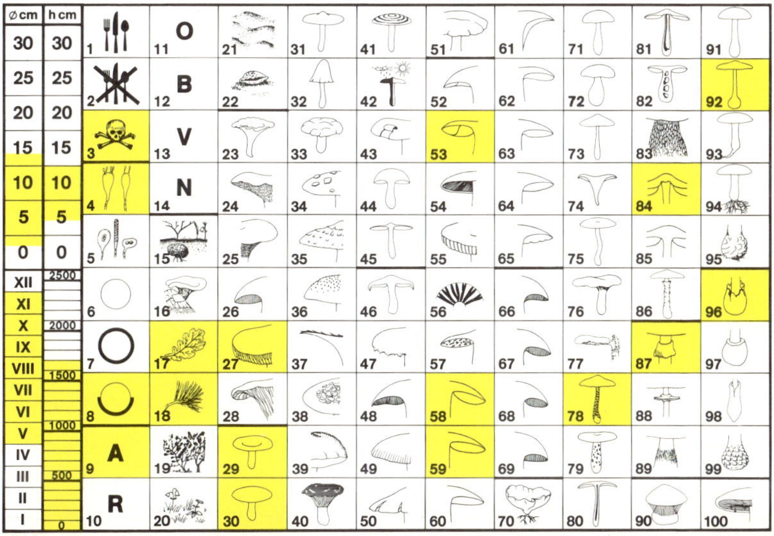

Stropharia aeruginosa Curtis ex Fries

Verdigris Agaric

Cap hemispherical, then plane, often umbonate; glutinous, then dry; bluish-green, turning yellow once picked; easily peeled. **Gills** ash-grey, then reddish-grey or purple-brown. **Stipe** concolorous with the cap, covered in whitish floccules; foot swollen and showing white mycelial cords; thin, fugacious ring, often blackened by the ripe spores. **Flesh** flaccid and whitish. **Smell** of turnip. **Taste** of turnip. **Spores** purplish-brown, 7·5–8·5 × 4·5–5 µ; oval. **Habitat** common on sandy soils and vegetal remains. **Edible/Poisonous** unworthy, although not poisonous, because of very poor taste.

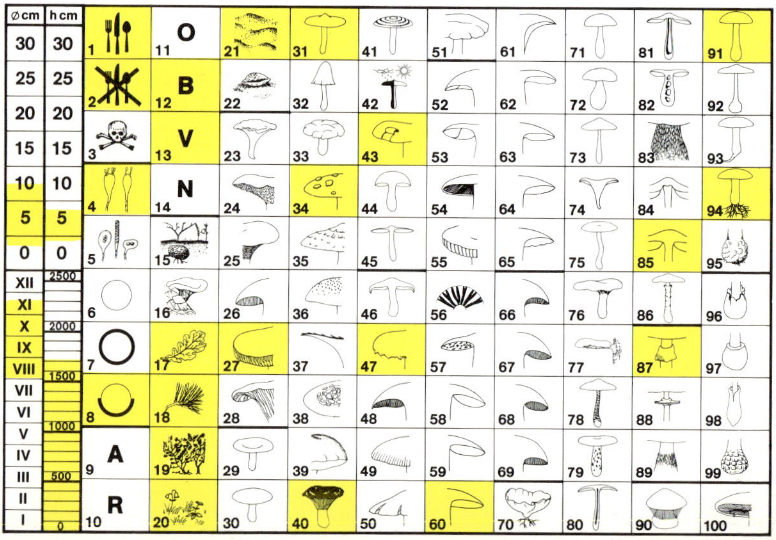

Cracked Russula or Green Russula
Cap convex, then plane or slightly depressed, often with involute margin; dry cuticle, yellowish-green or verdigris, rarely whitish; soon breaking into small polygonal plaques. **Gills** adnexed or almost free; white, then cream, flushed with pink. **Stipe** thick, often irregular, white; downy, full, then with marrow. **Flesh** firm, white. **Smell** pleasant. **Spores** white tinged with cream, 6–10 × 5–7 μ; elliptical, with irregular warts. **Habitat** under frondose trees, rarely under conifers, in numerous groups. **Edible/Poisonous** edible; of good quality even when raw.

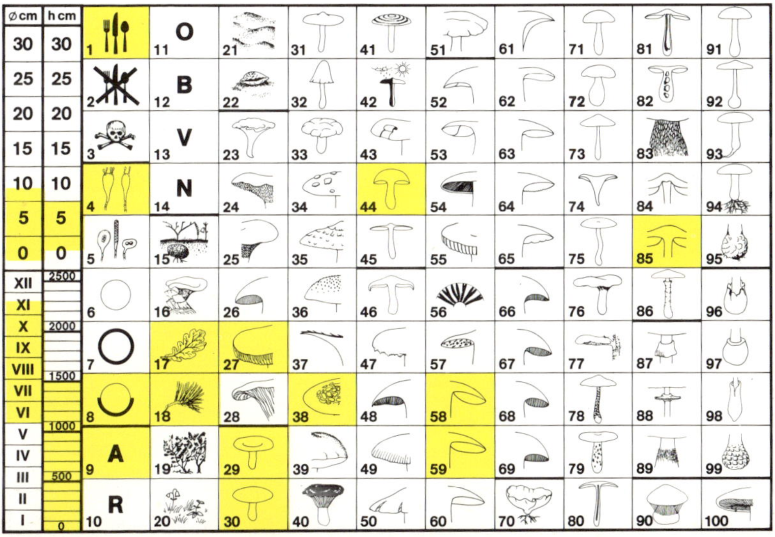

Russula olivacea Schaeffer ex Fries

Cap soon becomes plane then depressed; with smooth margin, occasionally wavy, later cleft; non-detachable cuticle, rough, of uneven and variable colour: olivaceous, olive-reddish, reddish-purplish; greyish-ochraceous. **Gills** crowded and thick, white tinged with cream; then almost ochraceous. **Stipe** stocky and club-shaped, rough, almost wrinkled; white tinged with rosy or spotted with reddish; full, then spongy. **Flesh** firm, fragile, white, slightly coloured under the cuticle. **Smell** almost nil. **Taste** of nuts, slight. **Spores** ochraceous, 7·5–10 × 7–9 μ; elliptical-oval with spikes. **Habitat** in the woods, in groups. **Edible/Poisonous** edible; of good quality when cooked.

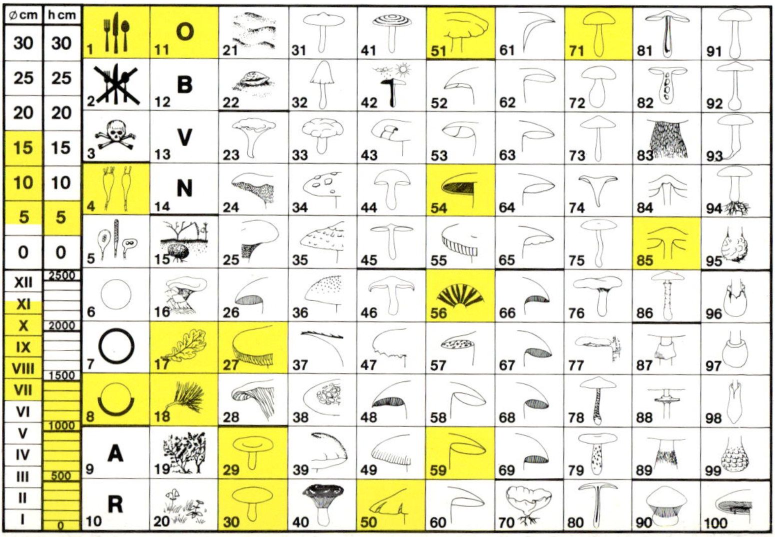

Syn. *Agaricus bitorquis* (Quél.) Sacc.
Cap plane to convex, white, tinged with ash-grey, fibrillose; becomes tinged with ochre. **Gills** pink, then brown or blackish, with a lighter edge. **Stipe** white then yellowish, characterised by two or three rings, the third less apparent. **Flesh** white tinged with pink. **Smell** pleasant. **Taste** pleasant. **Spores** chocolate brown, 4–6 μ; globular. **Habitat** on sunny soils rich in organic substances. **Edible/Poisonous** edible.

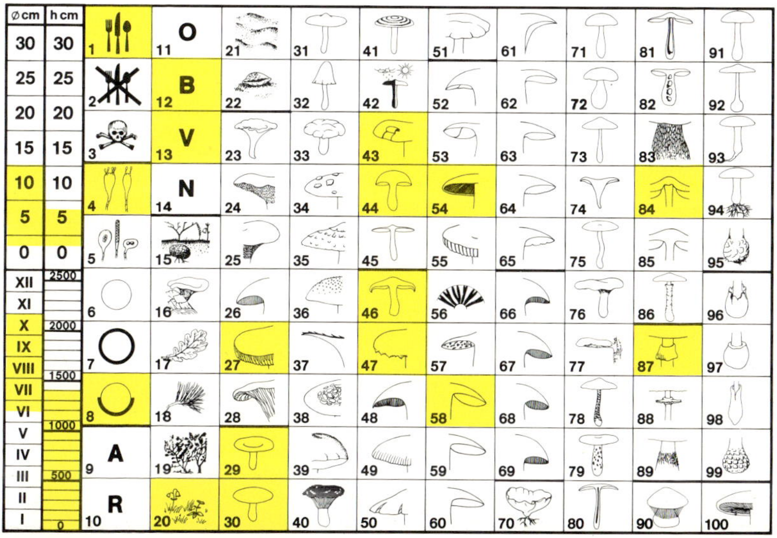

Field Mushroom

Syn. *Agaricus campester* (L.) Fr.

Cap plane to convex, white to ochraceous, tinged with pink in rainy weather. **Gills** first adnexed then free; white tinged with pink, then pinkish and finally brown or blackish. **Stipe** white and full. **Flesh** tender, white, slightly pink when exposed to the air. **Smell** of hazelnuts. **Taste** of hazelnuts. **Spores** cocoa-brown, 4·5–5·5 × 7–8·5 μ; oval. **Habitat** on soils rich in organic substances; often in circles of several individuals. **Edible/Poisonous** edible; of good quality.

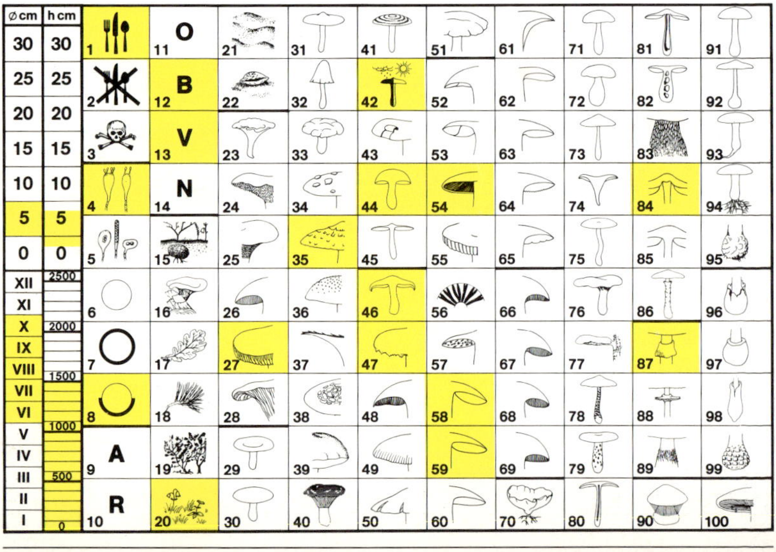

Yellow Stainer
Syn. *Agaricus xanthodermus* (Gen.)
Cap globular, then campanulate with plane top, finally more or less expanded or even slightly concave; whitish or greyish, darker in the centre; yellow when bruised; shiny like satin and often radially cracked. **Gills** ash-white, then pinkish and finally brownish-black. **Stipe** shiny white, yellow when bruised; almost bulbous foot, retaining mycelial fragments; white ring with yellow edge. **Flesh** white, but yellow in the foot. **Smell** of carbolic. **Taste** revolting. **Spores** brownish-black, 5–7 × 3–4 μ; oval. **Habitat** on rotting organic substances, in groups. **Edible/Poisonous** toxic; causes stomach and intestinal troubles.

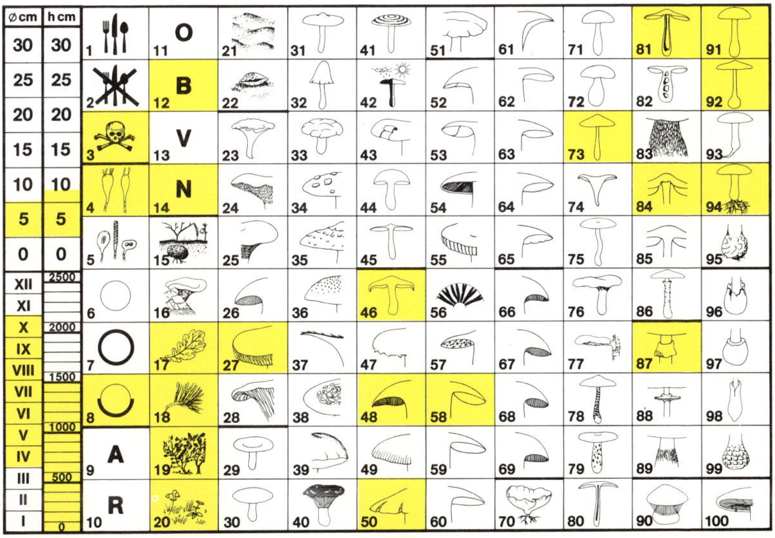

Straw Agaric
Syn. *Agaricus stramineus* (Schff. et Moell.) Sing.
Cap almost never completely expanded, with fringed margin; the surface is fibrillose to scaly, straw-coloured; in dry weather it breaks into regular fragments like truncated pyramids, separated by white furrows. **Gills** crowded, first adnate then free; whitish, then pinkish and finally blackish. **Stipe** fusoid, then almost cylindrical; wide, membraneous ring, white; full, then hollow. **Flesh** firm, white, pink when exposed to the air. **Smell** of sweet almonds. **Taste** of sweet almonds. **Spores** brownish-black, 10–12 × 6–6·5 μ; ovoid. **Habitat** in summer, in meadows, rarely in the woods, in groups, circles or tufts of few individuals. **Edible/Poisonous** edible; of good quality.

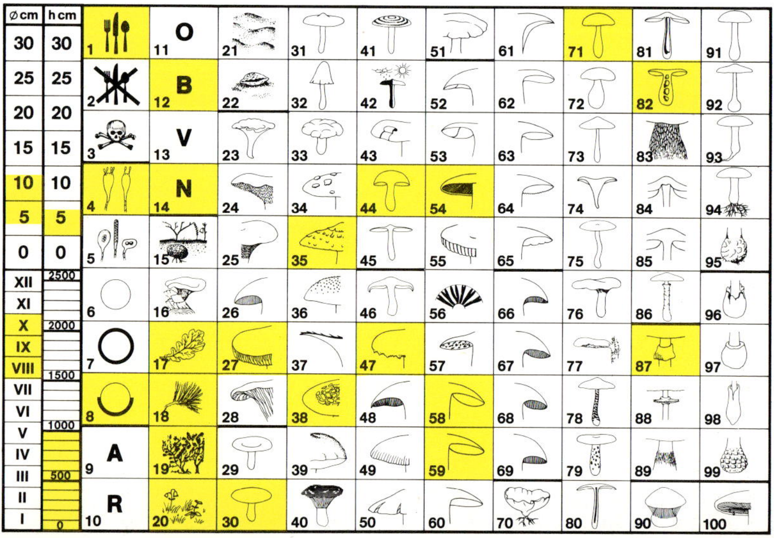

Shaggy Ink Cap or Lawyer's Wig
Cap ovoid, then campanulate; white, brownish on top; scaly cuticle down to the margin. **Gills** white, then purplish, finally turning into a black liquid. **Stipe** white, becoming grey when bruised; thin ring, mobile and fragile. **Flesh** tender in the cap, fibrous in the stipe. **Smell** pleasant. **Taste** pleasant. **Spores** black, 9–12 × 7–9 μ; elliptical. **Habitat** on soils with decomposing organic substances, particularly in the rainy season, in tufts, very common. **Edible/Poisonous** edible; of good quality; must be eaten within twenty-four hours as it deteriorates rapidly.

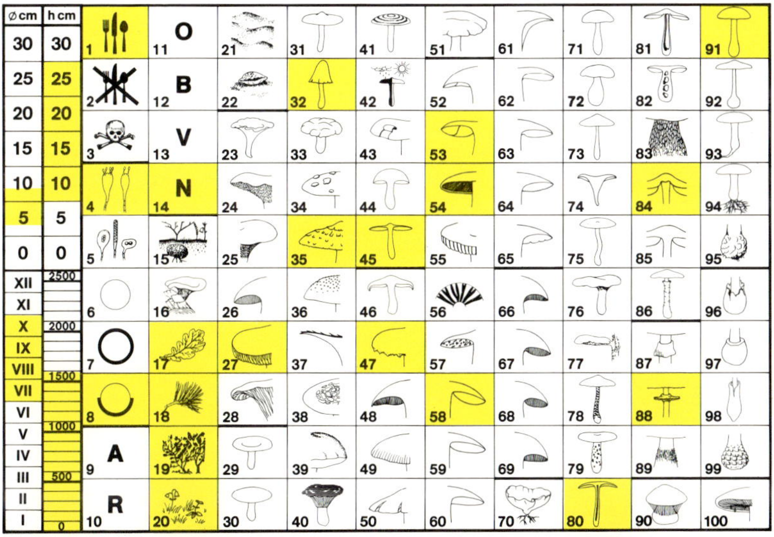

Cap slightly gibbous with protruding top; thin, overhanging margin, smooth, wavy and lobed; silky-white, tinged with ash or yellow, shiny in wet weather. **Gills** irregular edges, ash-white, then pink and finally tinged with red. **Stipe** the foot often joined to that of other individuals in the same tuft; fibrillose, silky-white. **Flesh** tender in the cap, slightly fibrous in the stipe; white, becoming pink when eaten by insects. **Smell** of flour. **Taste** of flour. **Spores** pink tinged with light brown, 8–10 × 7·5–9 μ; angular. **Habitat** in spring, at the foot of the *Rosaceae*, in tufts; often in great numbers. **Edible/Poisonous** edible; of good quality.

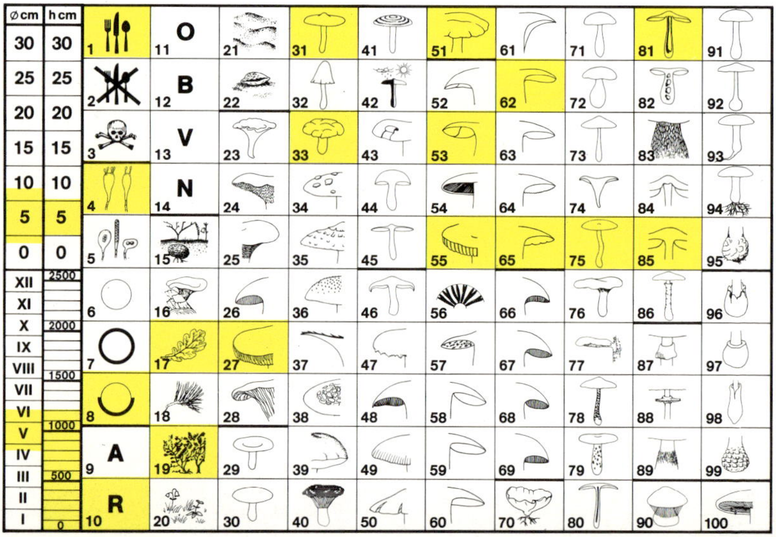

Syn. *Calocybe gambosa* (Fr.) Donk

Cap hemispherical, then expanded, with prominent top; margin often wavy; smooth; from light pinkish-brown to white. **Gills** white, then creamy. **Stipe** sometimes slightly fibrillose; white or tinged with cream; full. **Flesh** firm in the cap, even firmer in the stipe; white. **Smell** of flour. **Taste** of flour. **Spores** white, 5–6·5 × 3–4 μ; elliptical. **Habitat** usually in the mountains in spring, sometimes again at the end of summer; in groups. **Edible/Poisonous** edible; of good quality; can be preserved in various ways.

Clubfoot Clitocybe

Cap initially convex, with involute margin and slightly protruding top; then plane and depressed to become concave; finely tomentose; greyish-ochraceous with lighter margin, becoming greyish-brown in wet weather. **Gills** creamy, then yellowish. **Stipe** shaped like a club, the foot often enormously swollen, appearing pyriform; concolorous with the cap; spongy. **Flesh** flaccid, watery, whitish. **Smell** pleasant. **Taste** mild, slightly acid when soggy. **Spores** white, 5·5–7 × 3–4·5 μ; elliptical. **Habitat** in the woods, in small groups, not very common. **Edible/Poisonous** inedible.

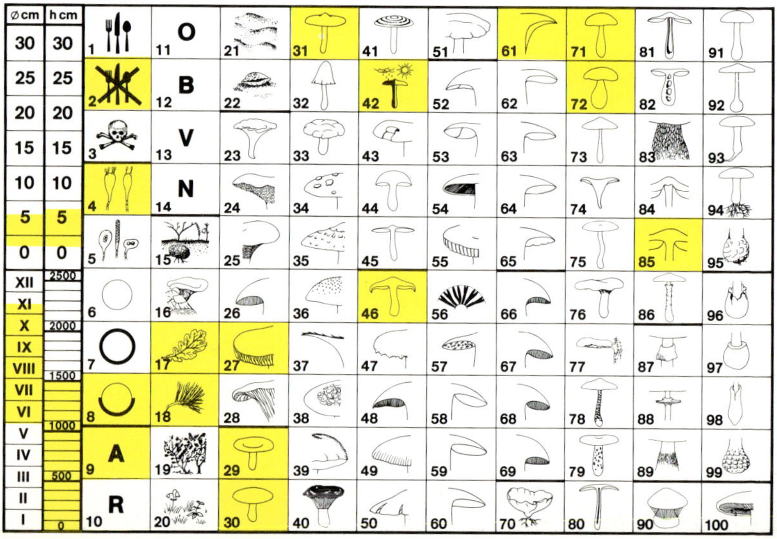

Giant Clitocybe
Syn. *Leucopaxillus giganteus* (Fr.) Sing.
Cap initially with involute margin, then open but extended beyond the gills; infundibuliform; smooth and scaly, creamy-white then much darker. **Gills** occasionally forked, whitish, then creamy, tinged with pink. **Stipe** fibrillose, white, then creamy or even darker; full. **Flesh** compact and firm in the cap; fibrous in the stipe; white. **Smell** mild. **Taste** pleasant. **Spores** white, 7–9 × 4–5 μ; elliptical with apex. **Habitat** in the mountains, meadows, in large circles. **Edible/Poisonous** edible, though mediocre.

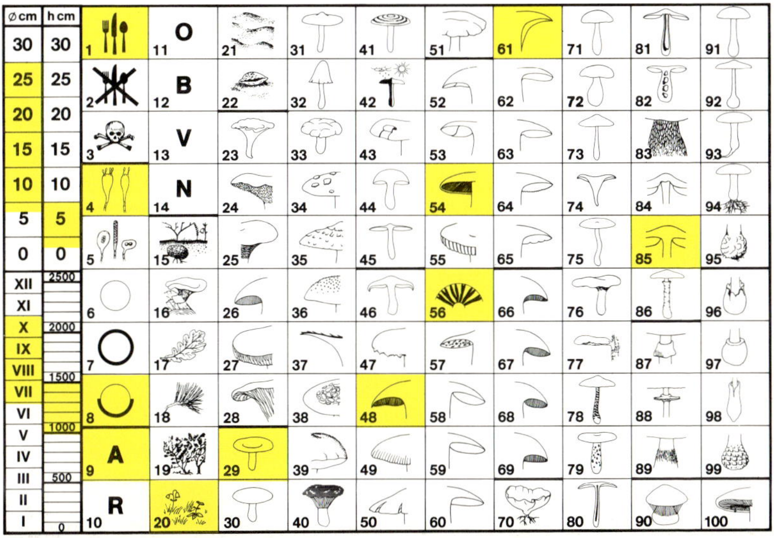

Cap hemispheric with involute margin; then expanded with overhanging margin; often gibbous, even depressed; smooth, opaque, ash-coloured, finally tinged with ochraceous and brownish tints. **Gills** decurrent, sparse, ash-white, then tinged with yellowish hues. **Stipe** fusiform, seldom centred; usually eccentric; white, full. **Flesh** compact in the cap, fibrous in the stipe; white. **Smell** of flour. **Taste** pleasant. **Spores** white, 10–13 × 5–6 μ; cylindrical with apex. **Habitat** usually in the mountains, in groups, on the remains of various species of *Umbelliferae*. **Edible/Poisonous** edible; of good quality.

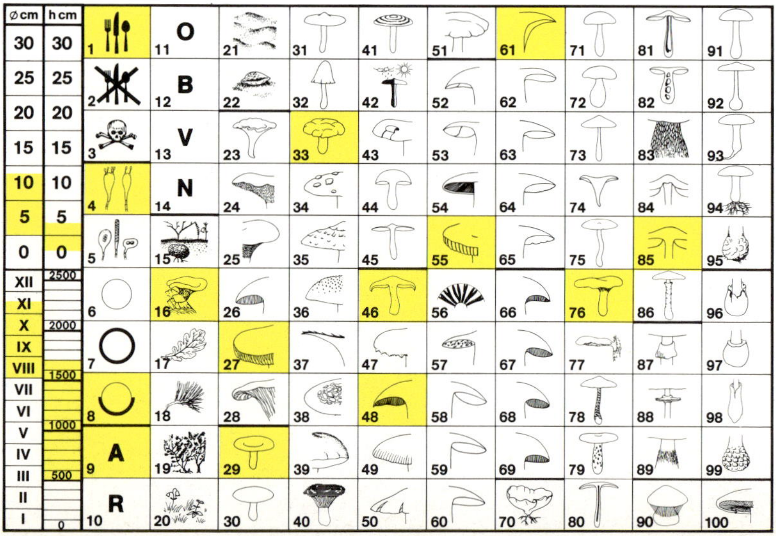

Cap with involute margin even when finally cup-shaped or infundibuliform; damp, white, with more or less concentrical, pinkish areas; often littered with earth and vegetal remains. **Gills** whitish then pinkish. **Stipe** thick, like an upside-down cone, sometimes eccentric; white, full. **Flesh** firm and white. **Smell** slightly disgusting. **Taste** acrid. **Milk** white, acrid. **Spores** white, sometimes tinged with pinkish-yellowish hues, 6–7 × 5–6 μ; globular. **Habitat** in late summer and until early winter, under poplars and chestnuts, rarely under other frondose trees; in groups of many individuals, often in circles. **Edible/Poisonous** inedible.

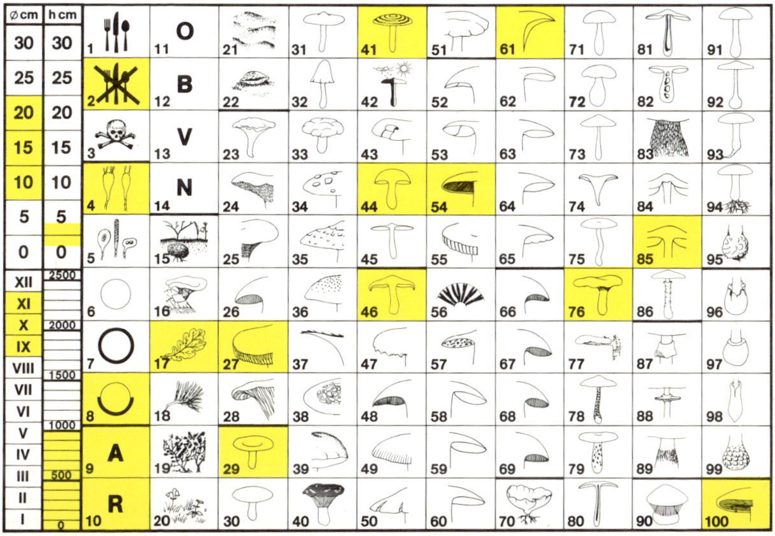

Hygrophorus poetarum Heim

Cap finally plane to umbonate; viscid in humid weather; scaly in the middle; from pinkish-cream to pinkish-ochre, darker in the middle; becomes a lighter colour with age; seldom totally white. **Gills** whitish then ivory. **Stipe** with pointed foot; white then yellowish; scaly; full. **Flesh** firm in the cap, fibrous in the stipe; white, but tinged with pink immediately under the cuticle. **Smell** of fruit. **Taste** slight. **Spores** white, 6–8·5 × 4–5 μ; oval. **Habitat** under beeches and oaks on high ground, rarely on plains; semi-circular, in groups. **Edible/Poisonous** edible; can be preserved in various ways.

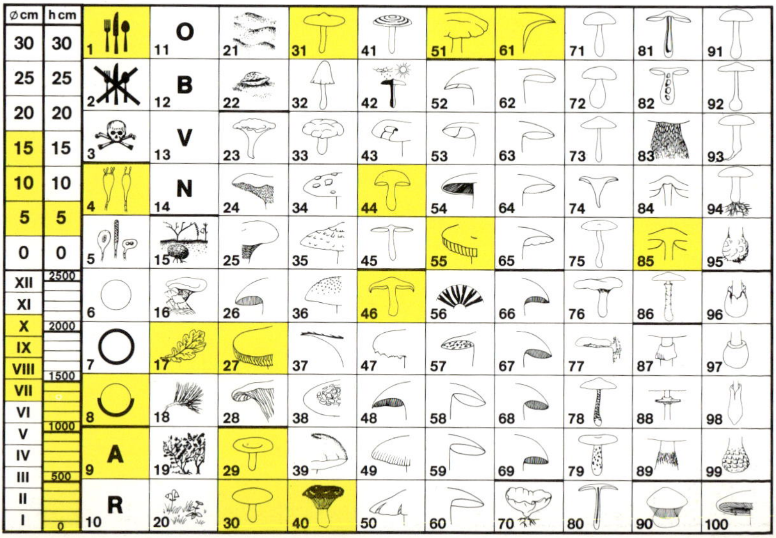

Cauliflower Fungus

It looks like a globular sponge, then a head of endive made of intersecting, creased ribbons, first white then yellowish. **Stipe** absent or very short; with mycelial cords at the base. **Flesh** somewhat resilient, elastic, fibrous, whitish and then ash-like. **Smell** pleasant. **Taste** slightly resinous. **Spores** ochraceous, 5–8 × 4–5 μ; elliptical to oval. **Habitat** in the autumn, among woody remains of conifers. **Edible/Poisonous** edible; of good quality.

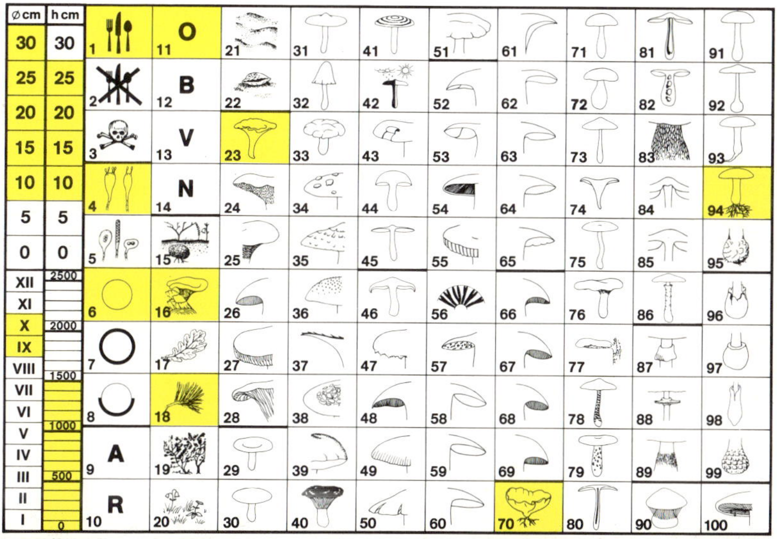

White Helvella
Cap like a saddle moving and curling unpredictably; the surface turning to the exterior, dry, yellowish-white; the underside ashy-brown. **Stipe** irregular, furrowed, often deeply; white, then whitish, or concolorous with the cap; hollow. **Flesh** fragile, white. **Smell** mild. **Taste** mild. **Spores** white, 18–22 × 9–13 μ; elliptical. **Habitat** in the autumn, in woods; rare. **Edible/Poisonous** edible.

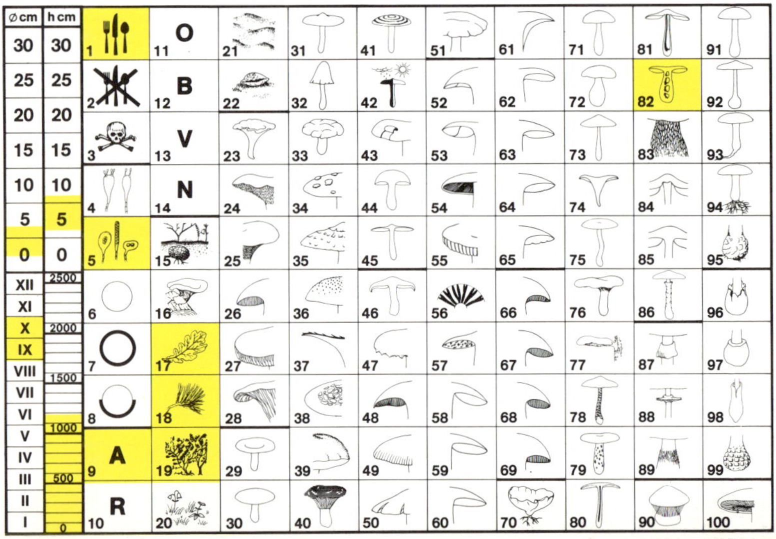

Syn. *Stropharia rugosoannulata* Farlow ex Murr.
Cap globular, then expanded; humid, with radial stripes, brownish-purple becoming lighter and scaly with age. **Gills** ash-coloured, then ash-purplish and finally sooty. **Stipe** white, tinged with yellowish, with fragile, star-shaped ring. **Flesh** doughy, white. **Smell** pleasant. **Taste** pleasant. **Spores** purplish-brown, 10–12 × 6·5–8 μ; elliptical to oval. **Habitat** on vegetal detritus, in groups or tufts; not so common. **Edible/Poisonous** edible.

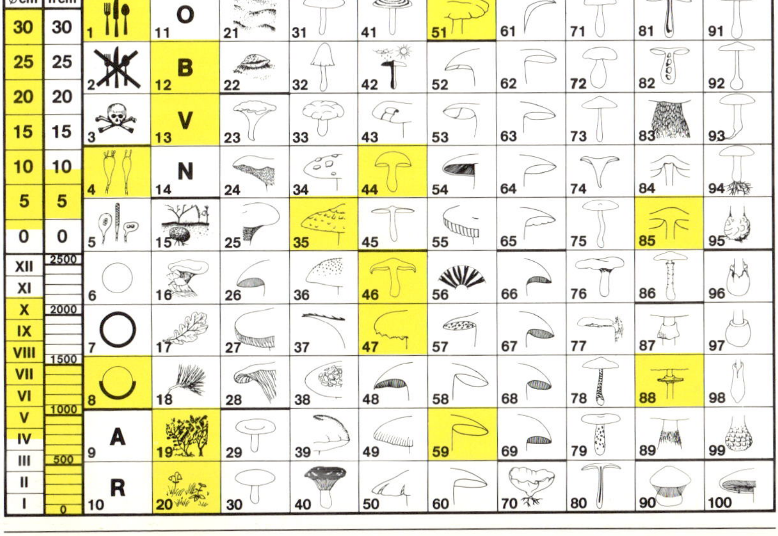

Violet Cortinarius

Cap globular or hemispherical, finally plane, with projecting centre; violet, velvety and fibrillose, scaly. **Gills** violet, then cinnamon. **Stipe** concolorous with the cap, sometimes with pseudo-ring left by the cortina; finally with marrow. **Flesh** purplish. **Smell** of cedar wood. **Taste** mild. **Spores** rust-coloured, 10–13 × 7–9 μ; elliptical. **Habitat** both under conifers and frondose trees, sometimes in groups of several individuals. **Edible/Poisonous** edible.

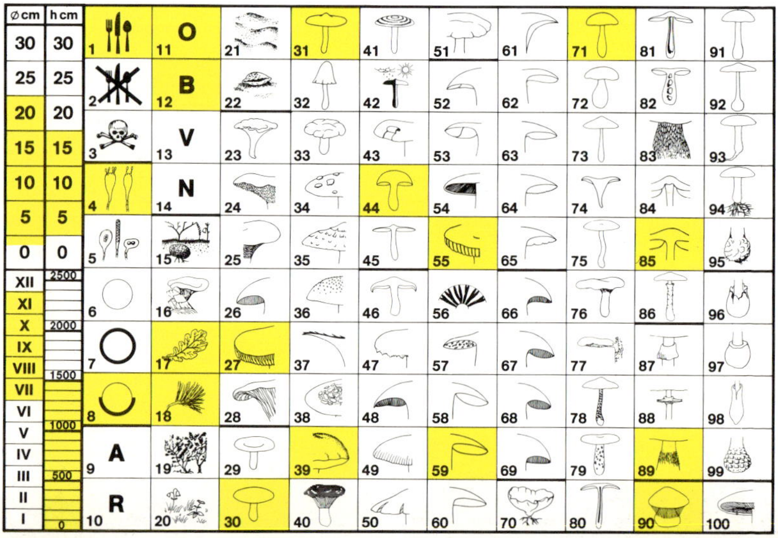

Cap hemispherical, then expanded, with involute and fringed margin; dry, fibrillose, shiny, violet, often with lighter floccules; with age it gets lighter and often becomes ochraceous. **Gills** purplish-white, then ochraceous and finally rusty, with paler edges. **Stipe** with fugacious, purplish remnant of the cortina; often with whitish floccules in a ring; full and somewhat spongy. **Flesh** firm in the cap, spongy in the stipe; saffron-yellow. **Smell** of acetylene. **Taste** bitter. **Spores** rusty, 8–9·5 × 5–5·5 μ; elliptical. **Habitat** under conifers, sometimes also under frondose trees. **Edible/Poisonous** inedible; could be poisonous.

Cap hemispherical, then considerably expanded; viscid in humid weather, from brown to purplish-brown, rusty with age; margin for a long time involute, with occasional light-blue floccules, then striate and sulcate. **Gills** whitish, then purplish-ochraceous. **Stipe** with swollen foot, but often ending in a point; whitish, with scaly fibrils, membraneous and purplish. **Flesh** firm, whitish in the cap, purplish-ochre in the foot. **Smell** faint. **Taste** pleasant. **Spores** rusty, 16–19 × 8–9 μ; amigdaliform. **Habitat** under frondose trees, sometimes also under conifers, on calcar, in groups. **Edible/Poisonous** edible; of good quality; can be preserved in various ways.

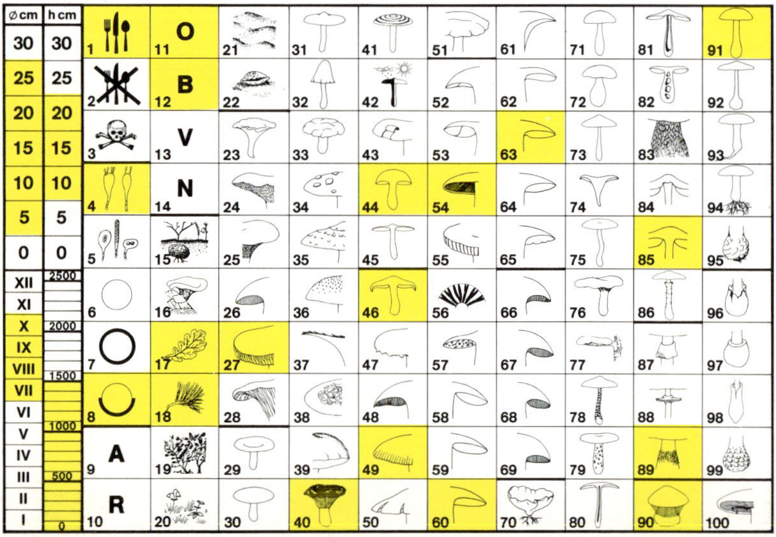

Clean Mycena
Cap with prominent top, then plane; striate margin; from pink to purplish, getting lighter in dry weather. **Gills** purplish-white. **Stipe** fibrillose, shiny, concolorous with the cap. **Flesh** juicy, fragile; purplish-white. **Smell** of turnip. **Spores** white, 5–8·5 × 2·5–4 μ; elliptical-cylindrical. **Habitat** almost everywhere in all seasons, sometimes in small groups. **Edible/Poisonous** not to be recommended, although not poisonous, as very poor.

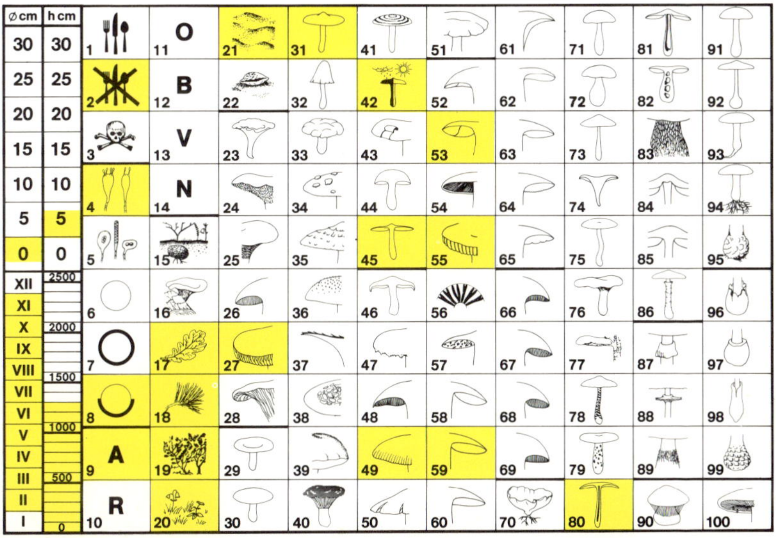

Tricholoma rutilans (Schaeffer ex Fries) Quélet

Purple Blewit

Cap sometimes with prominent top; covered in purplish-pink scales which give it a velvety feel; in time, it loses its scales, beginning with the margin, uncovering the yellow cuticle. **Gills** yellowish then yellow. **Stipe** slightly paler than the cap; fibrillose and sometimes scaly like the cap. **Flesh** whitish then yellowish. **Smell** faint. **Taste** slightly bitter. **Spores** white, 7–9 × 5–6 μ; elliptical-oval. **Habitat** on rotting trunks of conifers, occasionally in tufts. **Edible/Poisonous** not to be recommended, although not poisonous, as it remains bitter even after cooking.

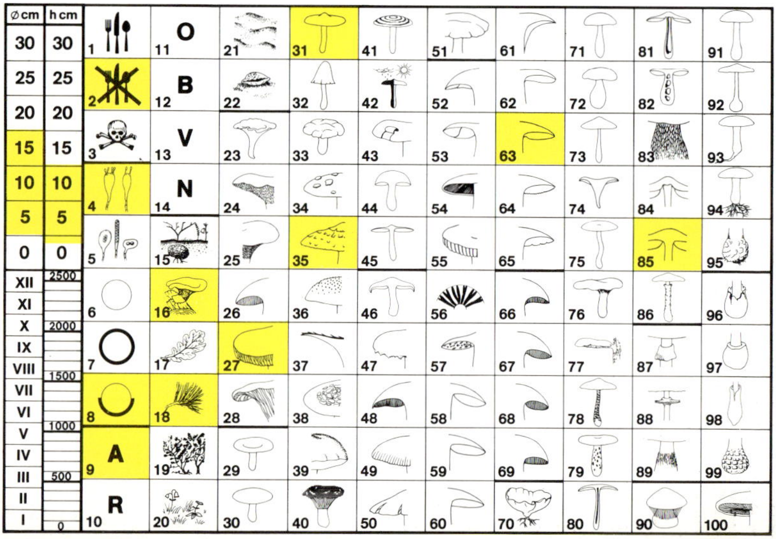

Tricholoma colossus Fries

Cap with a margin often wavy and lobed; easy to peel, slightly viscid, fibrillose, sometimes rough; reddish to brown tinged with purple. **Gills** whitish or yellowish, often with reddish blotches. **Stipe** thick with rudimental ring; concolorous with the cap, but paler at the top. **Flesh** firm, white, turning pinkish in the air. **Smell** almost imperceptible. **Taste** slightly bitter. **Spores** white, 7–10 × 5–7 μ; elliptical or oval. **Habitat** under conifers, rare, but abundant when established. **Edible/Poisonous** edible; of good quality.

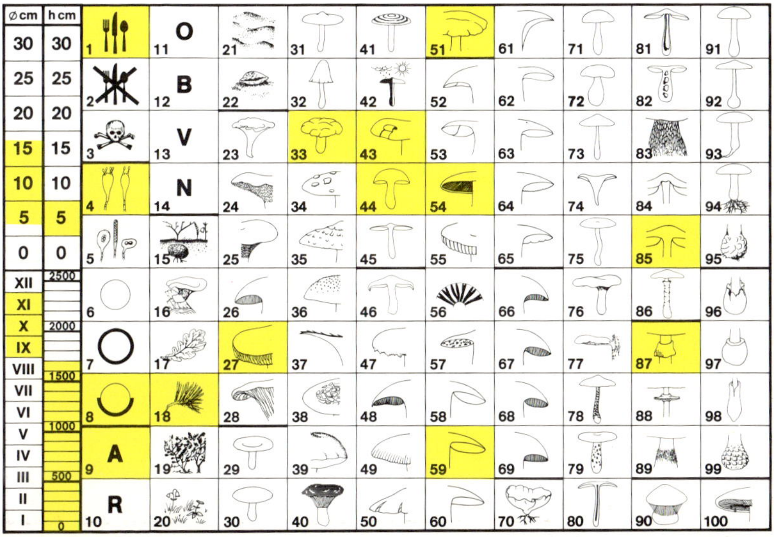

Wood Blewit or Naked Mushroom

Syn. *Lepista nuda* (Bull. ex Fr.) Cke. or *Tricholoma nudum* (Fr.) Kumm.

Cap finally often concave with wavy edge; from purplish to ochraceous. **Gills** very crowded, concolorous with the cap. **Stipe** fibrillose, with slightly enlarged, floccose foot; ash to purplish; full, then stringy. **Flesh** purplish. **Smell** aromatic. **Taste** pleasant. **Spores** purplish-pink, 6–8 × 4–5 μ; elliptical. **Habitat** in the autumn, sometimes in spring, under frondose or coniferous trees, in circular groups. **Edible/Poisonous** poisonous if raw; edible and of good quality when properly cooked.

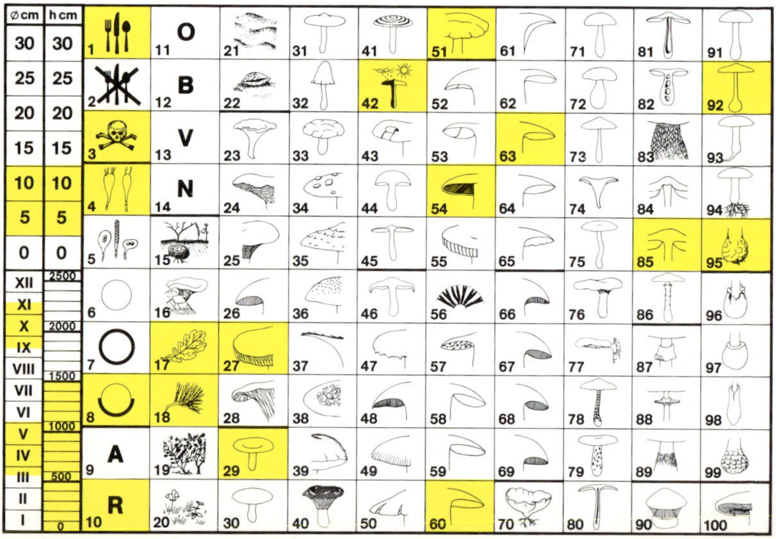

Charcoal Burner

Cap soon becomes plane, with wavy margin; viscid in wet weather; can be peeled almost in its entirety; blackish or greenish or purplish-grey or bluish or of intermediate hues. **Gills** waxy, white. **Stipe** pruinose, wrinkled; white, often tinged with brown or purple; full, then spongy. **Flesh** fragile and white, slightly coloured just under the cuticle. **Smell** faint. **Taste** of hazelnuts. **Spores** white, 7–10 × 5·5–6·5 μ; globular or elliptical, with tiny warts. **Habitat** in the woods, in groups. **Edible/Poisonous** edible; of good quality.

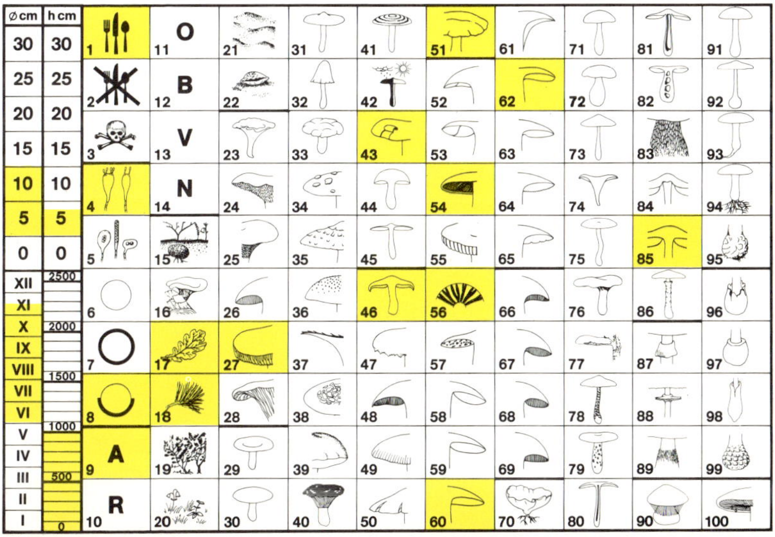

Gomphidius glutinosus Schaeffer ex Fries

Cap umbonate, smooth, glutinous, shiny, easy to peel, purplish-grey or violet-brown, blackening with age. **Gills** with slightly dented edge; decurrent; whitish, then sooty or blackish. **Stipe** sometimes swollen at the foot; glutinous and joined to the margin of the cap by a glutinous, purplish-white cortina; white, then showing the blackish-brown remains of the cortina; full. **Flesh** white, but a clear yellow at the foot. **Smell** faint. **Taste** faint. **Spores** sooty-brown or blackish, 18–21 × 5·5–7·5 μ; fusoid. **Habitat** under conifers in small groups. **Edible/Poisonous** edible, but the cuticle should be removed.

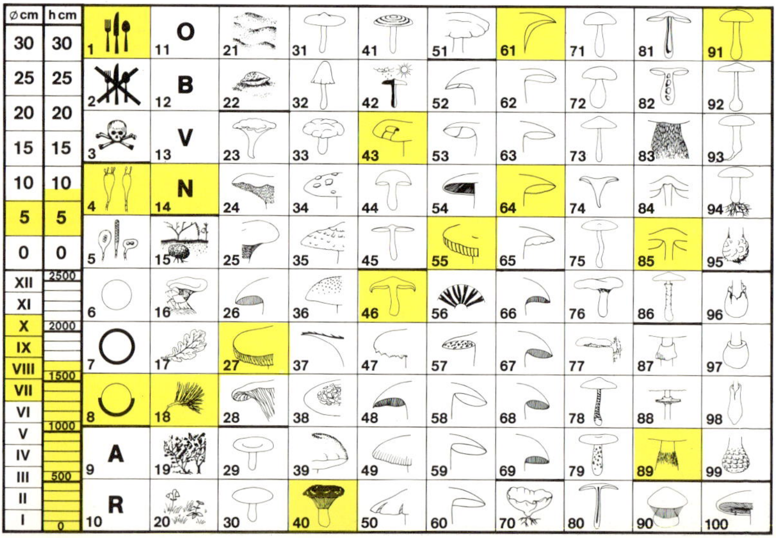

Cap hemispherical, then rounded with wavy margin; covered in light, whitish down which turns first blue then reddish or pinkish at the slightest touch; purplish-black patches appear with age. **Tubes** 2–3cm long; first adnexed, then free; yellowish-olivaceous. **Pores** small, round, irregular, yellow but soon red or orange; turning deep blue when touched. **Stipe** pear-shaped, then cylindrical with swollen foot; yellow, with reticle initially yellow then red; fading with age, when the reticle turns purplish; becomes deep blue when bruised; full. **Flesh** firm but soon flaccid; yellow, turning deep blue in the air. **Smell** musty. **Taste** acid. **Spores** olive-brown, 11–15 × 5–6 μ; fusoid. **Habitat** under beeches and oaks. **Edible/Poisonous** poisonous.

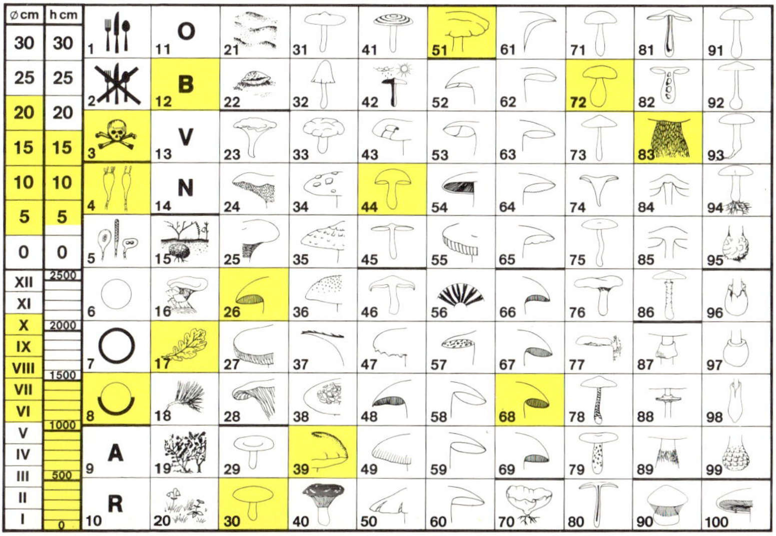

Shining Ganoderma

Cap fan-shaped, often irregular, with concentric areas protruding slightly; shiny, reddish-brown or violet-brown or reddish-violet. **Tubes** 0·5–0·8cm long, thin and brownish. **Pores** small, round, white then brownish. **Stipe** lateral, often irregular, smooth and shiny, concolorous with the cap or sometimes darker, almost blackish; full. **Flesh** elastic and juicy, but soon becoming woody. **Smell** non-existent. **Taste** non-existent. **Spores** brown, 8–13 × 6–9 μ; oval with apex and warts. **Habitat** at the foot of several plants. **Edible/Poisonous** inedible.

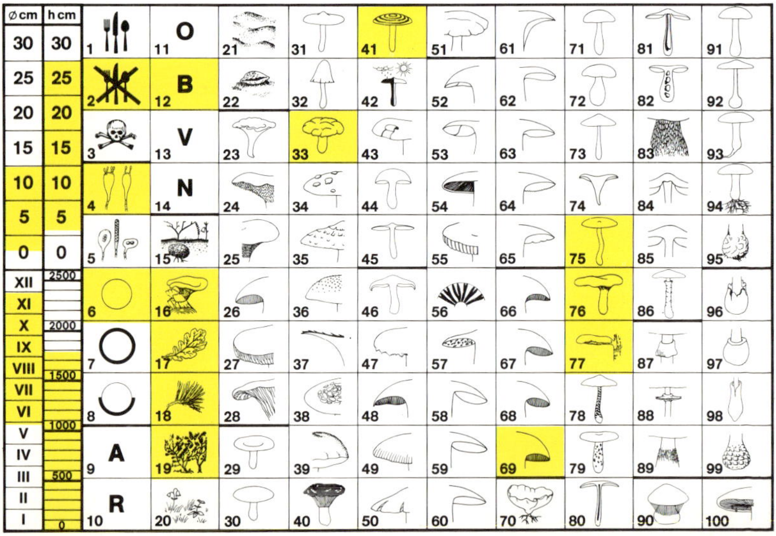

Clustered Chanterelle or Pig's Ear
Syn. *Gomphus clavatus* (Fr.) S. F. Gray
Club-shaped or turbinate, depressed at apex, sometimes almost funnel-shaped; occasionally open on one side; lobed and wrinkled margin; violaceous above, then nut-brown; paler along the side; marked longitudinally by veins which occasionally merge and reach almost to the base; the foot is white and pruinose. **Flesh** firm, white, shaded with pink at the edge. **Smell** mild. **Taste** mild, but acrid to bitter when old. **Spores** yellowish, 10–14 × 4–6 μ; elliptical. **Habitat** in the woods, rare, but abundant where established. **Edible/Poisonous** edible.

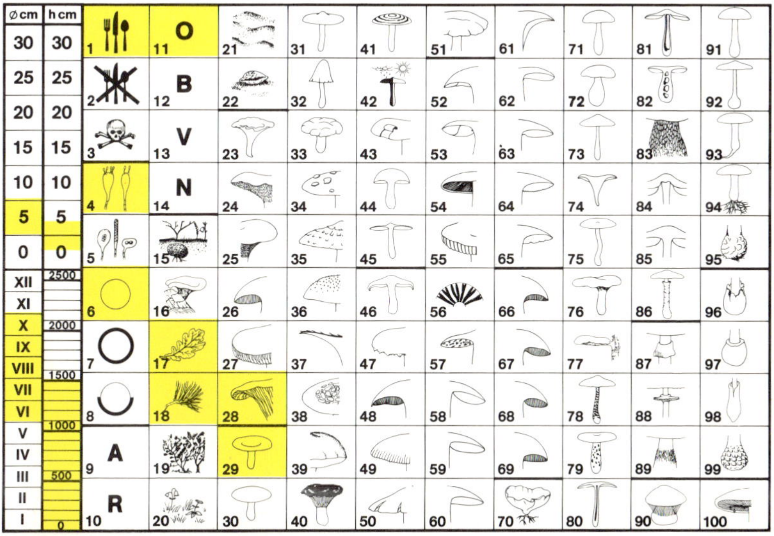

Craterellus cornucopioides Linnaeus ex Fries

Horn of Plenty or Trumpet of the Dead
Shaped like an empty cornucopia with lobed margin; the internal surface is scaly and sooty; the exterior smooth or slightly wrinkled, ash-grey to grey; turning black in wet weather. **Flesh** membraneous, elastic, tough, sooty. **Smell** pleasant. **Taste** pleasant. **Spores** white, 12–15 × 7–10 μ; elliptical. **Habitat** in woods, particularly coniferous; in large groups. **Edible/Poisonous** edible; of good quality; can be preserved by drying.

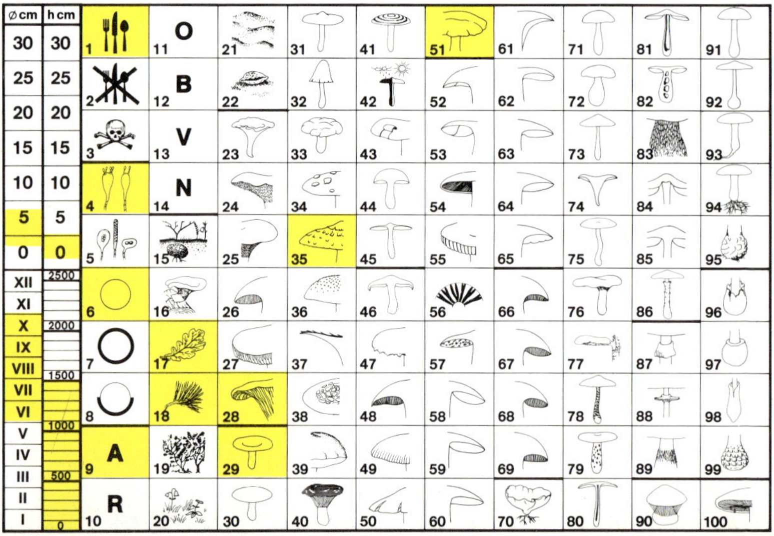

Purple Stinkhorn
First enclosed by a thick volva and steeped in yellowish mucilage, it looks like a small egg (3–4·5cm diameter), violaceous; the the stipe emerges from the open volva: it is 10–12 × 3–4cm, cylindrical, fistulous, spongy, whitish; it bears an ovoid cap (4–5 × 3–4cm) with a round hole in the apex, a thin collar round the hole, covered in a greenish mucilage which disappears on maturity leaving a whitish, honeycombed surface. **Flesh** white, porous, fragile, disgustingly smelly. **Spores** greenish-yellow, 3–4 × 2–2·5 μ; elliptical. **Habitat** on soils rich in organic matter. **Edible/Poisonous** inedible.

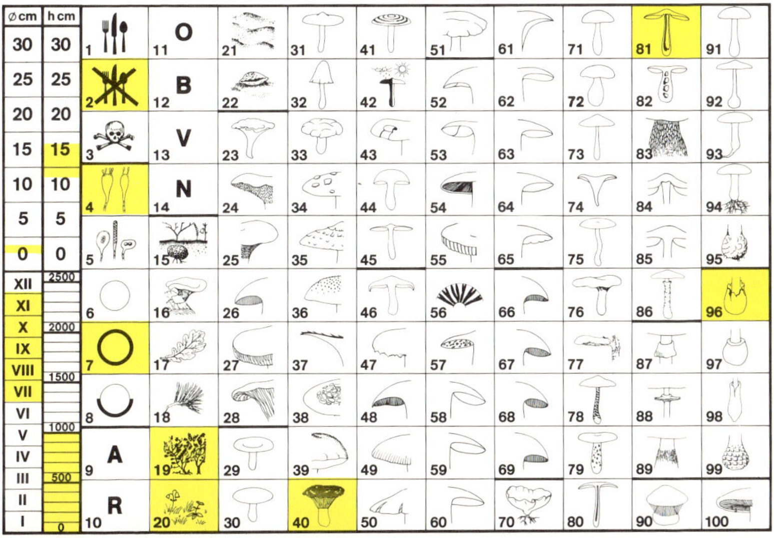

Lysurus mokusin (L.) Fries

A membraneous volva surrounds the base of a short stipe, pentagonal or hexagonal in section, orangey-pink, crowned by five or six tentacles which are eventually covered with a greenish mucilage. **Flesh** spongy to honeycombed, fragile, whitish, extremely smelly, sweetish. **Spores** greenish, 3.5–5 × 1·8–2·2 μ; elliptical. **Habitat** on woody remains or woodshavings. **Edible/Poisonous** edible.

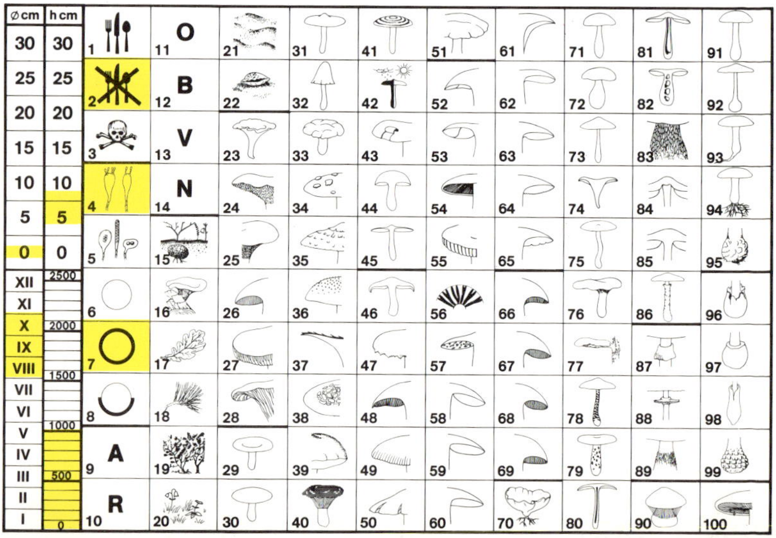

Syn. *Calvatia cyathiformis* (Bosc.) Morgan
Globular or ovoid; whitish, then ochraceous and finally brownish-violaceous; wrapped in a double peridium, thick and strong; as it ripens it becomes floppy and wrinkled. **Flesh** white, then tinged with violaceous. **Smell** pleasant. **Taste** pleasant. **Spores** brownish-violaceous, 6–10 μ; globular. **Habitat** both on low and high ground, in open positions. **Edible/Poisonous** edible; of good quality.

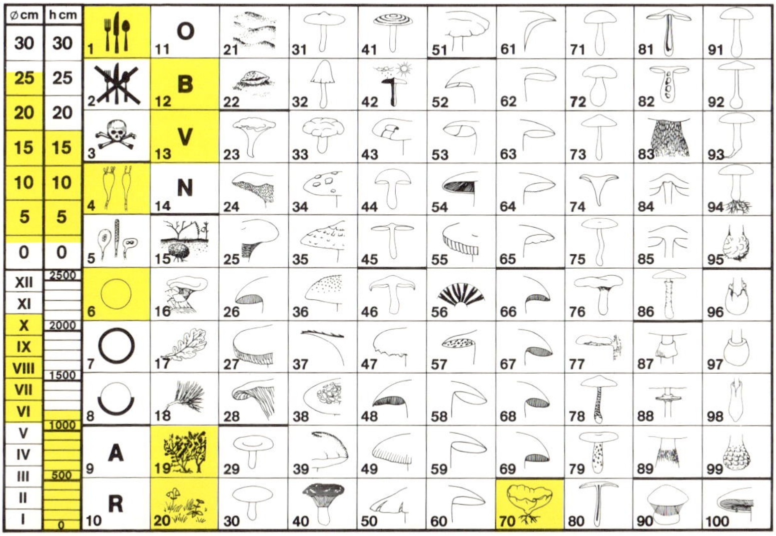

Cap almost globular, the colour of yellow narcissi, sometimes faded; viscid, warty, rarely glabrous; the margin soon striate; later expanded, dry, with a darker centre. **Gills** white. **Stipe** white, fibrillose; white ring, soon disappearing; bulbous foot; white, closely-adhering volva with neat rim, sometimes represented by a few fibrils and floccules. **Flesh** white, but yellowish under the cuticle of the cap. **Smell** either undetectable or very slight; of turnip. **Taste** pleasant. **Spores** white, 9–12 × 7–9 μ; elliptical. **Habitat** usually on sandy or calcareous soils. **Edible/Poisonous** edible; of good quality.

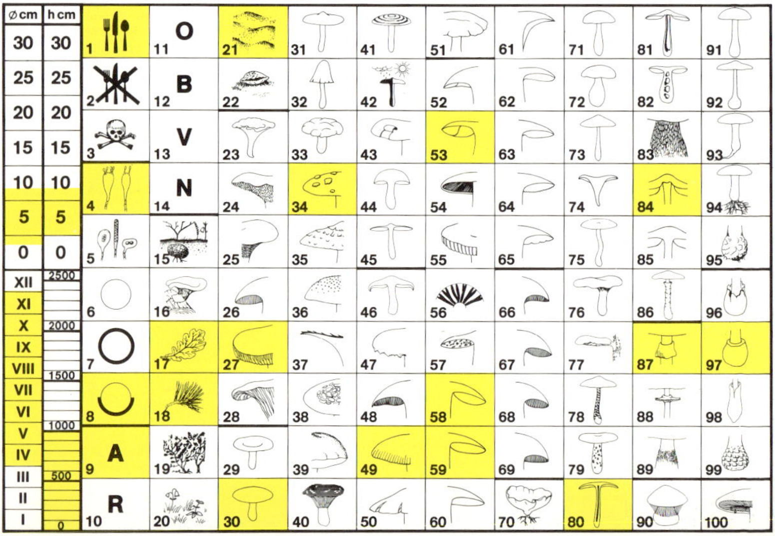

False Death Cap

Cap globular, then expanded and sometimes even depressed while remaining prominent in the centre; yellow, but white in the variety *alba*; with white, flocculo-fibrillose warts. **Gills** white, then whitish. **Stipe** white, then yellowish; floccose volva, with clear, step-like edge, ash-grey; striate and yellowish ring. **Flesh** white, soon becomes damp. **Smell** of turnip. **Taste** slightly acrid. **Spores** white, 8–11 × 7·5–9 μ; elliptical to round. **Habitat** on sandy soils, light and loose; common. **Edible/Poisonous** to be avoided, although edible, as it closely resembles the deadly *Amanita phalloides*.

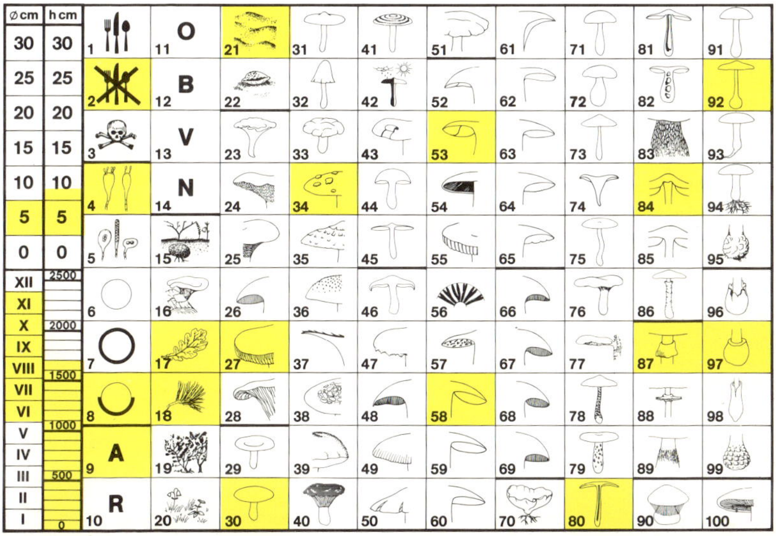

Sulphur Tuft or Clustered Wood-lover
Syn. *Hypholoma fasciculare* (Huds. ex Fr.) Humm.
Cap hemispherical, then plane, with prominent apex; margin with deciduous fringes; yellow to orange, darker on the apex. **Gills** yellowish, then olivaceous and finally brownish. **Stipe** curved, fibrillose, slightly paler than the cap; with pseudo-ring left by the cortina. **Flesh** tender in the cap, fibrous in the stipe; yellowish. **Smell** unpleasant. **Taste** bitter. **Spores** brownish-violaceous or blackish, 6–7·5 × 4–5·5 μ; elliptical. **Habitat** in all seasons, except the coldest part of winter, on dead or live wood; in tufts. **Edible/Poisonous** inedible, as it remains bitter even when cooked.

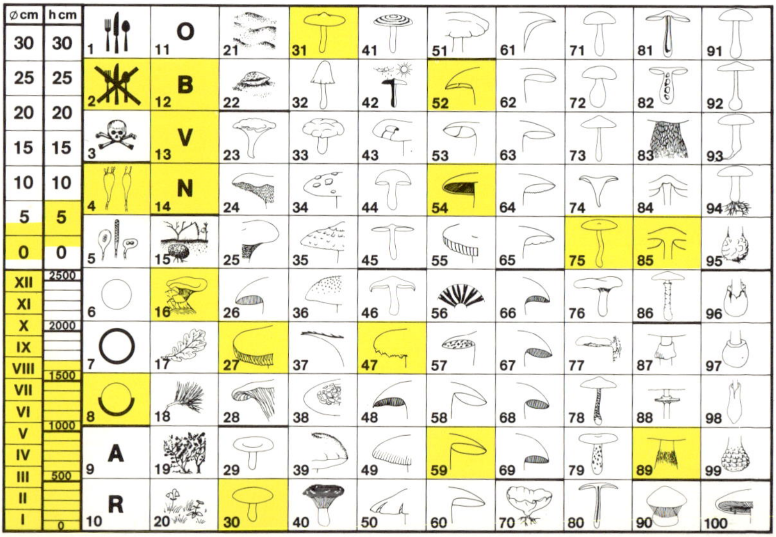

Shaggy Pholiota or Scaly Cluster Fungus

Cap hemispherical, then more or less expanded, with festooned margin; nut-brown, covered with brown scales, paling with age. **Gills** sulphur-yellow, then rusty. **Stipe** concolorous with the cap, scaly, with soft ring and a foot joined to that of other individuals in the same tuft. **Flesh** firm in the cap, fibrous in the stipe; white but shaded with yellow in the foot. **Smell** faint. **Taste** acid or of turnip. **Spores** nut-brown, 6–8·5 × 4–5 μ; elliptical. **Habitat** in the autumn (although not exclusively) on stumps and trunks of frondose trees; in tufts. **Edible/Poisonous** not recommended, although not poisonous, as taste unpleasant.

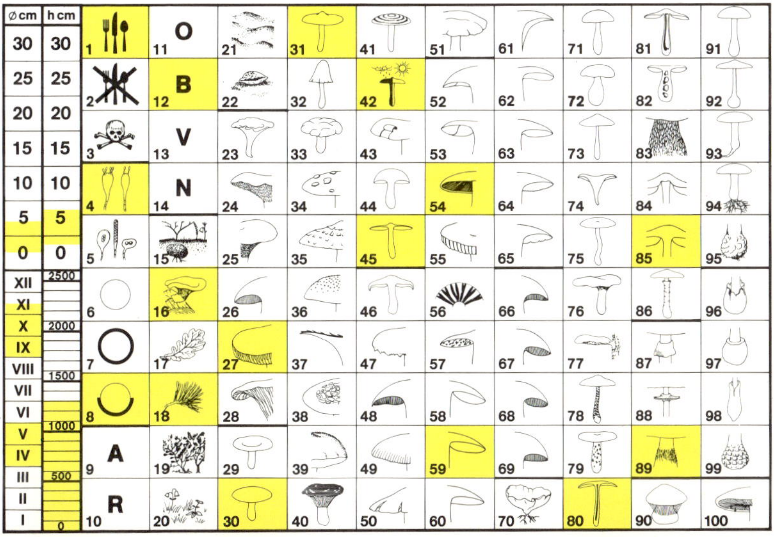

Syn. *Hypholoma capnoides* (Fr. ex Fr.) Kumm

Cap hemispherical, then more or less expanded but remaining in general rather cuspidate; wavy margin with fringes; yellowish, darker in the centre; shiny in wet weather. **Gills** ash-yellowish, then greyish-violaceous and brownish-grey. **Stipe** slightly paler than the cap, with brownish foot; generally with pseudo-ring left by the cortina. **Flesh** yellowish in the cap, brownish in the foot. **Smell** mild. **Taste** pleasant. **Spores** brownish, 7–9·5 × 3·5–5 μ; elliptical-oblong. **Habitat** more frequently in the autumn but sometimes also in spring on the stump of conifers; in tufts. **Edible/Poisonous** edible.

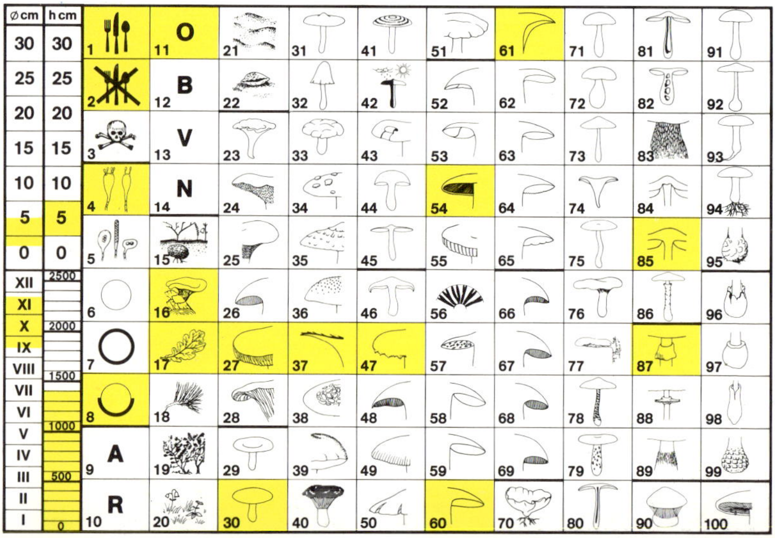

Changing Pholiota
Syn. *Kuehneromyces mutabilis* (Schff. ex Fr.) Sing. & Smith
Cap convex, then plane, sometimes scaly at first; yellowish-brownish, darker in wet weather. **Gills** briefly decurrent; whitish then nut-brown or cinnamon. **Stipe** often curved, scaly, with fragile and fugacious ring; yellowish from the ring upwards, brownish downwards. **Flesh** tinged with brown. **Smell** pleasant. **Taste** pleasant. **Spores** rusty-brown, 6–7 × 3–5 μ; ovoid. **Habitat** from spring to autumn on tree-trunks; in tufts. **Edible/Poisonous** edible.

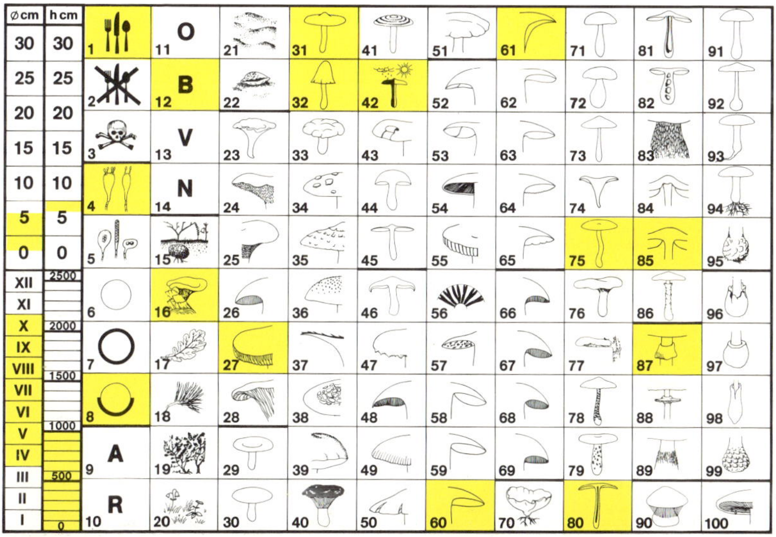

Collybia fusipes Bulliard ex Fries

Spindle Shank

Cap with thin, smooth margin; rusty-brown, turning pale and cracked in dry weather. **Gills** forked, with wavy and dented edge, white or creamy with rusty spots. **Stipe** wrinkled, more or less concolorous with the cap; stringy and hollow. **Flesh** elastic and strong, white. **Smell** pleasant. **Taste** bitter after much chewing. **Spores** white, 4–5·5 × 3–4 μ; elliptical. **Habitat** from spring to autumn, on the stumps of frondose trees. **Edible/Poisonous** edible, but poor, and only when young and without the stipe.

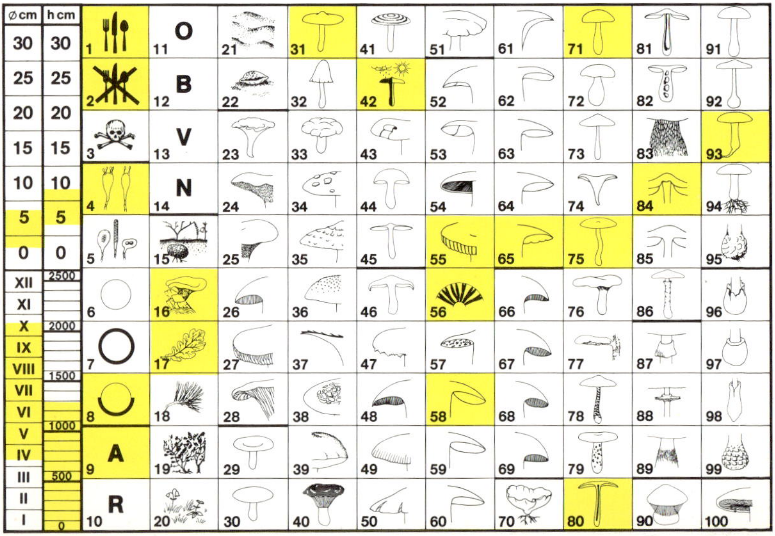

Winter Fungus or Velvet Shank
Syn. *Flammulina velutipes* (Curt. ex Fr.) Sing.
Cap orange-yellow, paler at the margin; viscid in wet weather; margin finally striate. **Gills** yellow. **Stipe** concolorous with the cap immediately beneath it; the rest black and velvety. **Flesh** rather firm in the cap, fibrous and tough in the stipe; yellowish. **Smell** pleasant. **Taste** mild or slightly bitter. **Spores** white 8–10 × 3–4 μ; elliptical-oblong. **Habitat** from September to March, during the winter if the weather is not too cold; on any kind of wood, in tufts. **Edible/Poisonous** edible; of good quality, except for the stipes which are rather tough.

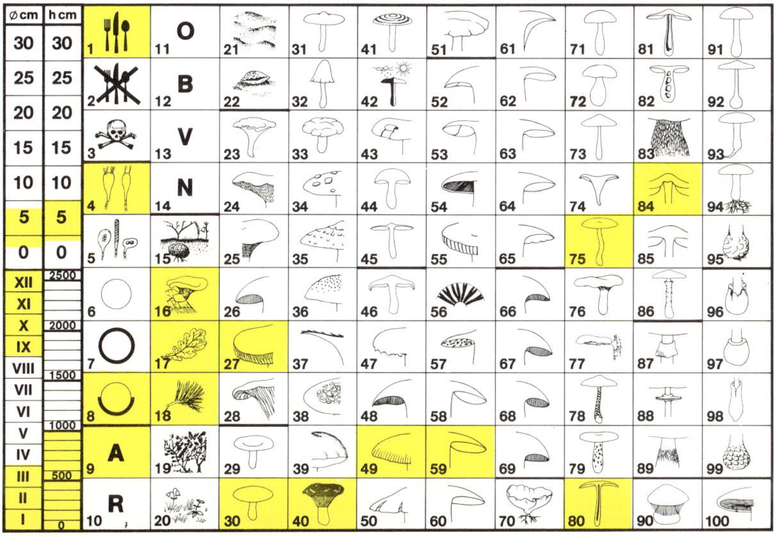

Yellow Knight Fungus

Syn. *Tricholoma flavovirens* (Pers. ex Fr.) Lund. & Nannf.

Cap convex, then plane, often slightly irregular; dry; yellow or tinged with green, with fuscous scales particularly in the middle. **Gills** with dented edge; yellowish, with greenish hues, then yellow. **Stipe** firm, smooth, concolorous with the cap; full. **Flesh** rather firm; white. **Smell** faint. **Taste** pleasant. **Spores** white, 6–8 × 4–5 μ; elliptical. **Habitat** in the woods, up to early winter. **Edible/Poisonous** edible; of good quality.

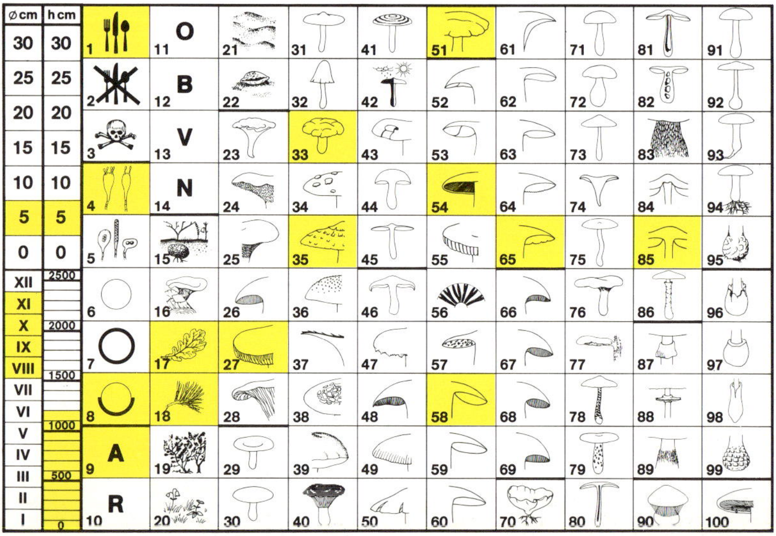

Cap generally with prominent apex; drooping and wavy margin; glabrous, with tiny radial fibrils; damp in the middle; from yellow to orange. **Gills** white, then yellow, with darker edge. **Stipe** white, fibrillose, lemon-yellow towards the end; full. **Flesh** generous, firm, white; yellowish towards the margin. **Smell** faint, of flour. **Taste** slightly bitter. **Spores** white, 4·5–6·2 × 3·5–4·2 μ; elliptical-ovoid. **Habitat** in the mountains, in summer, under conifers. **Edible/Poisonous** edibility unknown.

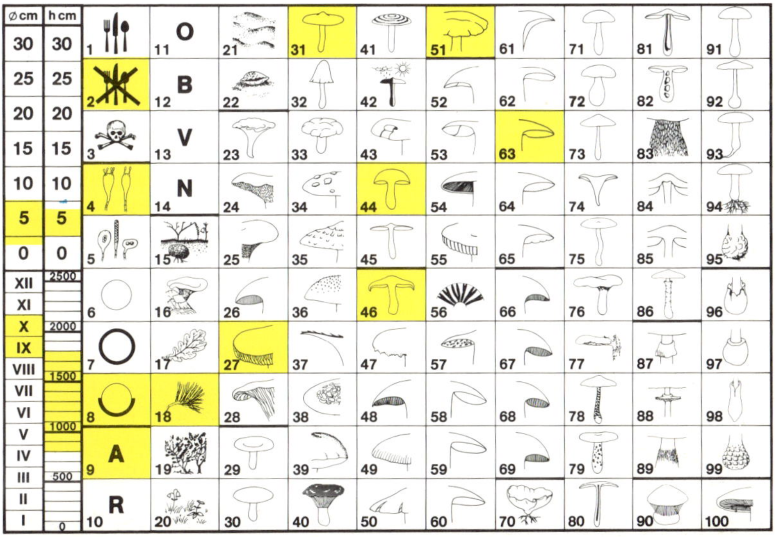

Tricholoma sulphureum Bulliard ex Fries

Sulphur Tricholoma

Cap hemispherical with involute margin; then expanded and umbonate; dry, fibrillose or glabrous; sulphur-yellow, sometimes brownish in the middle. **Gills** concolorous with the cap. **Stipe** sometimes slightly wavy; fibrillose, concolorous with the cap. **Flesh** firm, yellowish. **Smell** of acetylene or soap; unpleasant. **Taste** disgusting. **Spores** white, 9–12 × 5–6 μ; elliptical. **Habitat** in the woods, in groups. **Edible/Poisonous** inedible.

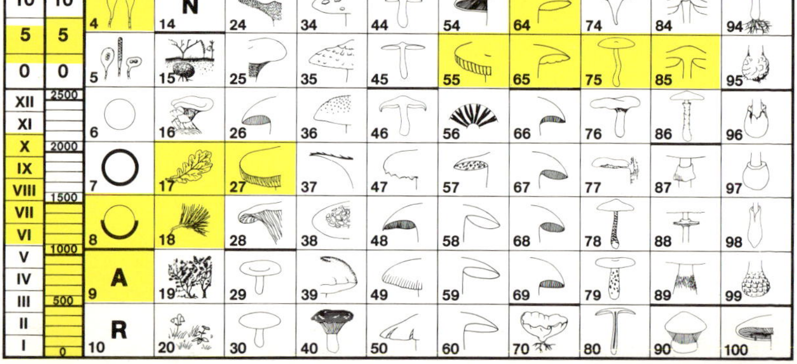

Honey Fungus or Boot-lace Fungus

Cap from white to sooty brown, often yellow, orange or brown depending on the plants it grows on; with scales or granules on the prominent apex, but sometimes totally glabrous. **Gills** whitish, then yellowish, with rusty patches. **Stipe** fibrillose, slightly lighter than the cap; wide ring, white then yellowish; full, then fistulous. **Flesh** firm in the cap, fibrous and tough in the stipe; whitish. **Smell** faint and pleasant. **Taste** bitter and slightly acrid; astringent. **Spores** white, 7–9 × 5–6 μ; oval-oblong. **Habitat** on various kinds of wood, in tufts, often quite large. **Edible/Poisonous** poisonous when raw; edible when cooked, as long as it is not too old.

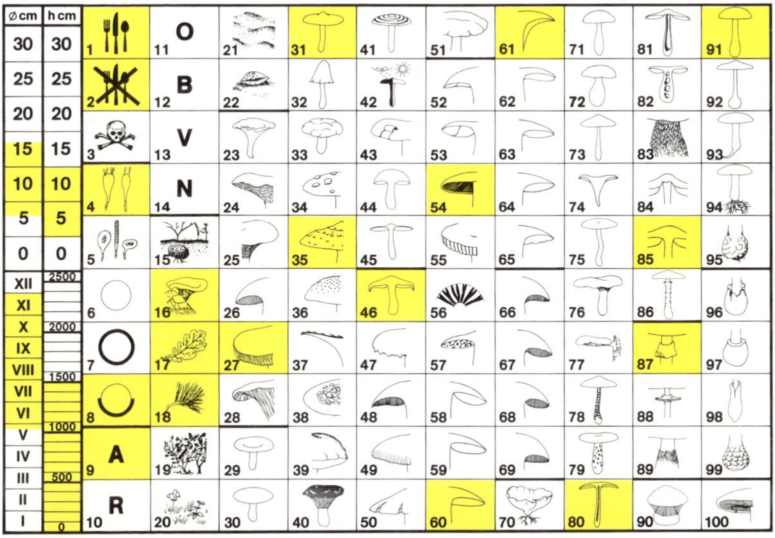

Cap with fringed margin which remains involute even when the cap becomes cup- or funnel-shaped; damp, with fine scales; yellow or tinged with ochraceous, sometimes with darker concentric areas. **Gills** whitish then pinkish-cream. **Stipe** stumpy, concolorous with cap, pitted, then hollow. **Flesh** firm, yellowish. **Smell** light; of fruit. **Taste** acrid. **Milk** plentiful, acrid, white; quickly turning sulphur-yellow when exposed to the air. **Spores** creamy, $6–7 \times 5–6\ \mu$; elliptical-globular. **Habitat** in the woods, in large groups. **Edible/Poisonous** inedible.

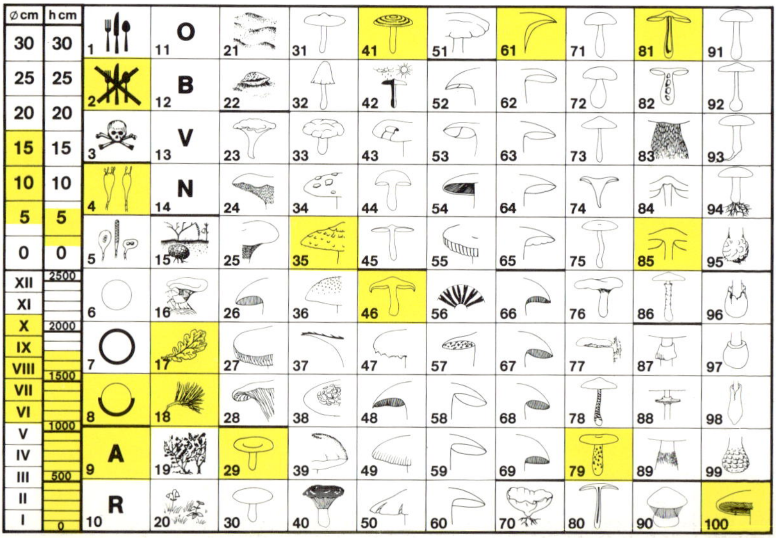

Geranium-scented Russula

Cap convex, slightly concave towards the end; often striate margin; viscid in wet weather, otherwise dry; ochraceous-yellow all over. **Gills** crowded, white, fragile. **Stipe** often a little swollen at the foot and tapering towards the cap; white or yellowish; with a chalky skin, spongy inside. **Flesh** whitish. **Smell** almost nil. **Taste** bitter. **Spores** tinged with cream, 7·5–9 × 6–7 μ; elliptical-globular. **Habitat** mainly under beeches; rather common. **Edible/Poisonous** inedible.

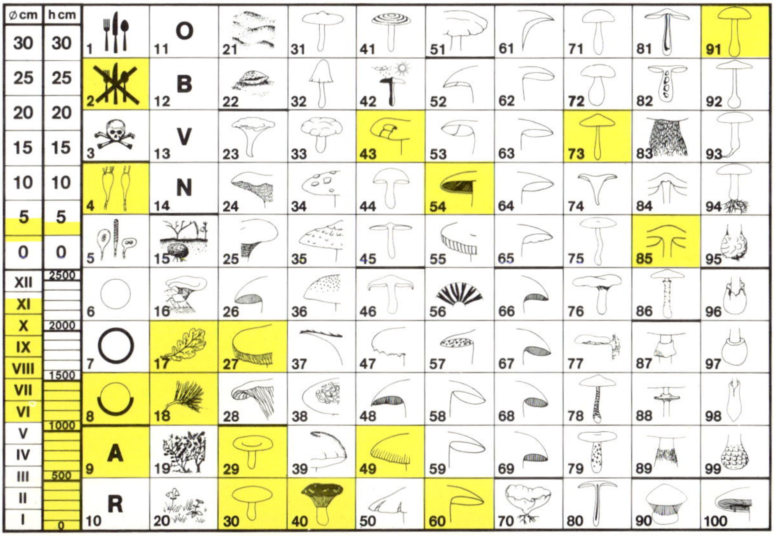

Russula foetens Persoon ex Fries

Stinking Russula

Cap slightly depressed towards the end; margin deeply striate; very viscid in wet weather; brownish. **Gills** white, with droplets, then with rusty spots. **Stipe** whitish, then yellowish with brownish patches; primrose; full, then hollow. **Flesh** fragile, creamy, acrid. **Smell** foetid. **Taste** acrid. **Spores** pale cream, 7–10–5 μ; globular. **Habitat** in the woods, in groups or scattered. **Edible/Poisonous** inedible.

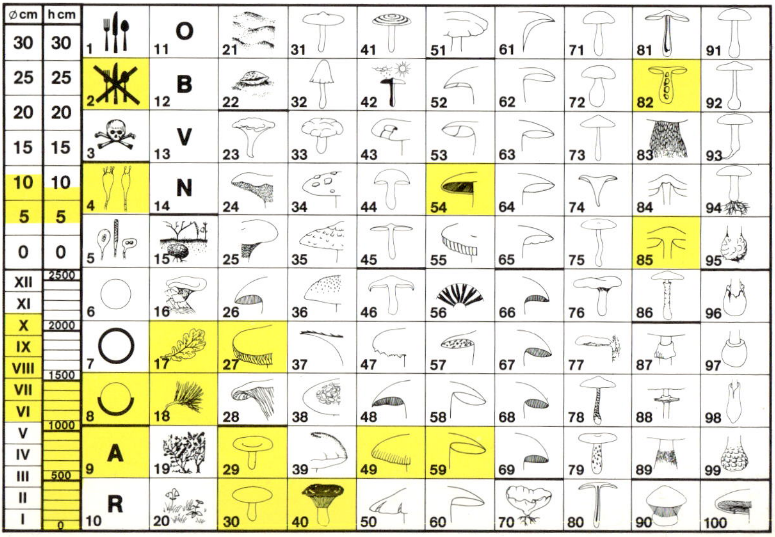

Meadow Wax Cap or Buff Cap or Butter Mushroom

Cap dry, with wavy margin; pinkish-ochraceous, darker in the centre where it cracks easily; with age it turns pale nut-brown all over. **Gills** cream. **Stipe** sometimes with pointed foot, sometimes tapering towards the cap; slightly lighter than the cap; full. **Flesh** firm in the cap, fibrous in the stipe; white tinged with cream. **Smell** pleasant. **Taste** pleasant. **Spores** white, 6–8 × 4·5–6 μ; oval. **Habitat** mainly in the mountains, but not exclusively. **Edible/Poisonous** edible; of good quality; can be preserved in various ways.

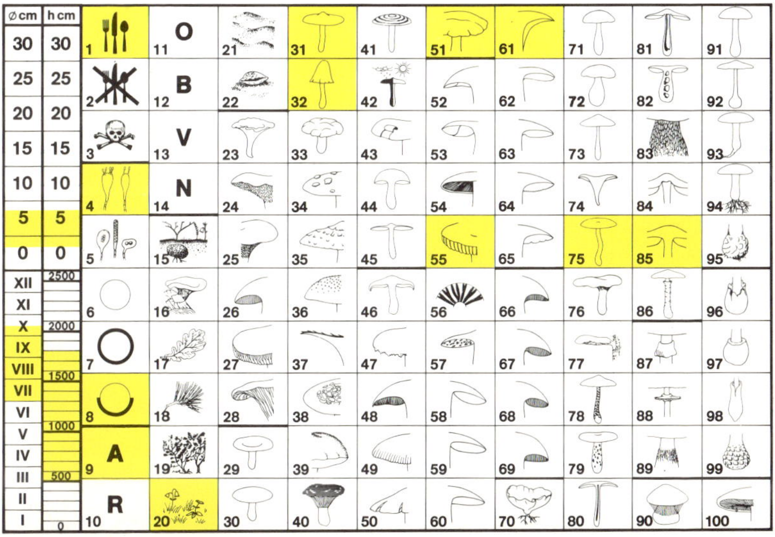

Viscid Bolete
Syn. *Boletus aeruginascens* Secr.
Cap with involute and festooned margin; then plane, often with prominent apex; ash-grey with various tinges, often spotted; viscid, easily peeled. **Tubes** 1–2·5cm long, adnate and sometimes slightly decurrent; ash-grey. **Pores** angular; whitish then ash-grey, rusty at the end. **Stipe** often with swollen foot; floccose, viscid, white with areas of different colours; whitish, fugacious ring; full. **Flesh** flaccid in the cap, fibrous in the stipe; whitish, but yellowish in the foot; turning bluish when exposed to the air. **Smell** pleasant. **Taste** pleasant. **Spores** brownish, 8–13 × 4–5 µ; fusoid. **Habitat** under larches. **Edible/Poisonous** edible.

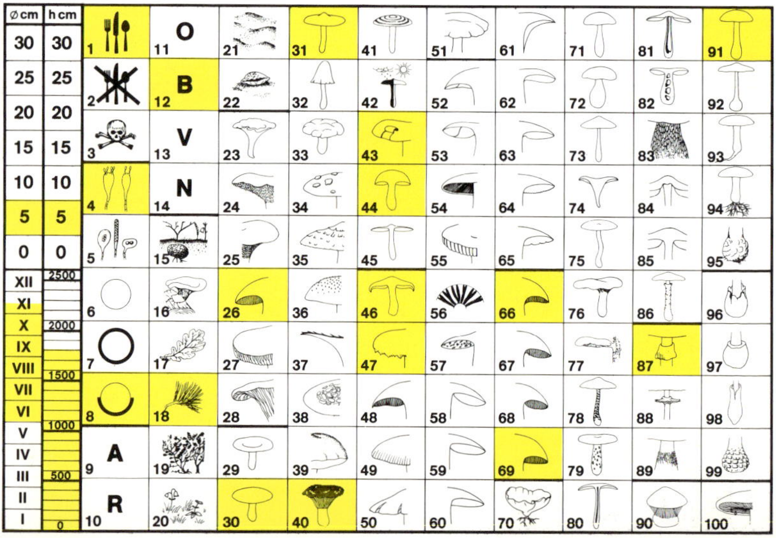

Chicken of the Woods or Sulphur Polypore

Caps variously shaped, gibbous; extending horizontally one above the other, becoming thinner at the edges with wavy or lobed margin; they join together in the interior forming a single mass or a single stipe, thick, whitish then yellowish; on the upper surface the colour varies from yellow to orange, sometimes with pinkish areas; when old they become ochraceous. **Tubes** short, 0·3–0·5cm; entirely covering the underside of the cap; sulphur-yellow. **Pores** miniscule; white tinged with sulphur-yellow, exuding yellowish dew. **Flesh** tender and yellow, becoming hard and woody; pale. **Smell** strong and pleasant in the fresh specimen, mild when dried. **Taste** acrid-bitter. **Spores** white-sulphur, then white, 5–6·5 × 4–5 μ; elliptical. **Habitat** from spring to autumn, on the stump of various trees. **Edible/Poisonous** inedible.

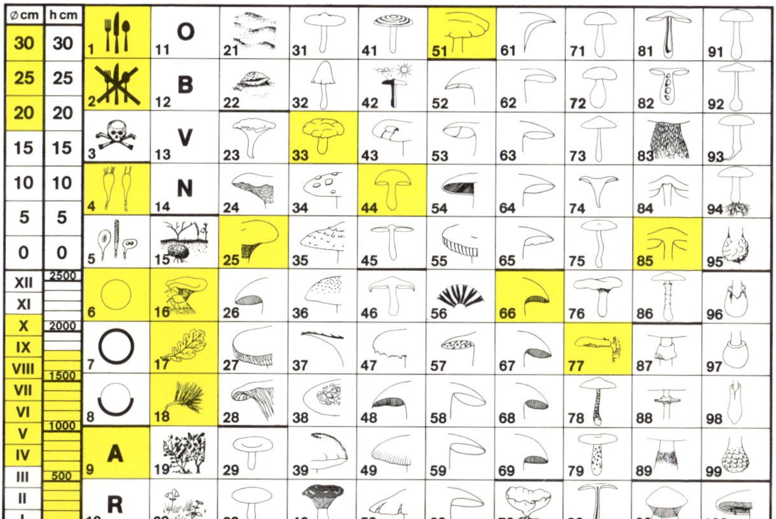

Yellow Clavaria
Resembling a cauliflower, with stocky points, hardly dented; at first a sulphur-yellow, then tinged with orange and finally brownish; fleshy stipes growing from a trunk, white or yellowish then brownish. **Flesh** fragile and white. **Smell** pleasant. **Taste** pleasant. **Spores** tinged with yellow, 9–12 × 4–5 μ; elliptical. **Habitat** in the woods. **Edible/Poisonous** edible while very young; inedible when old.

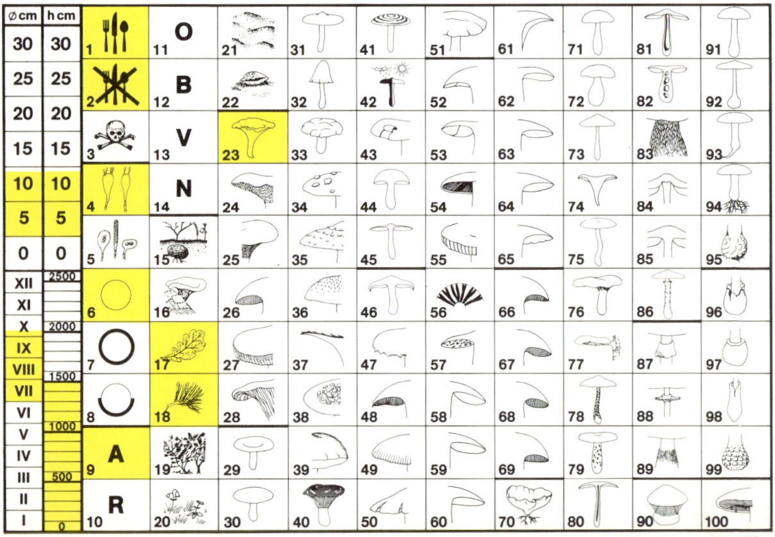

Chanterelle
Cap very irregular, with wavy and lobed margin; glabrous, from yellow to orange. **Pseudo-gills** rough and distant, occasionally joined; concolorous with cap; decurrent. **Stipe** tapering towards the foot; concolorous with cap; full. **Flesh** firm, slightly fibrous, white or yellowish. **Smell** mild. **Taste** pleasant, but bitterish in specimens grown under fir trees. **Spores** yellowish, 7–10 × 4–5·5 μ; elliptical. **Habitat** in the woods from May to November; in groups. **Edible/Poisonous** edible; of good quality; can be preserved in various ways.

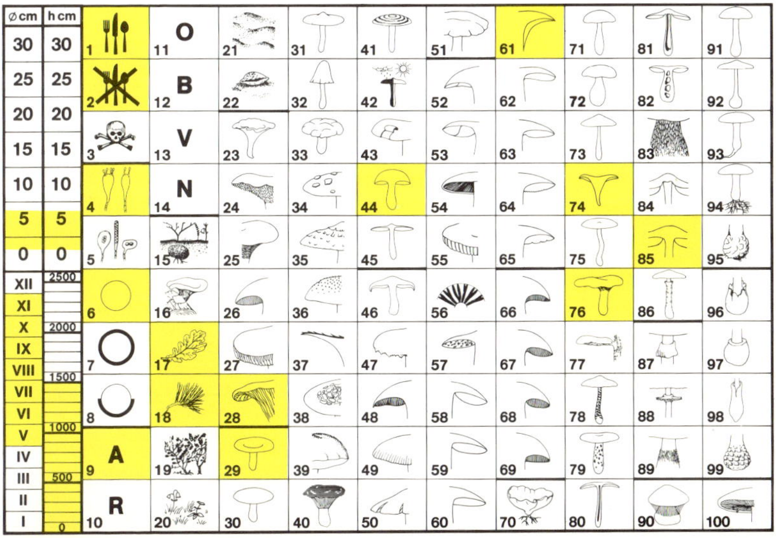

Cantharellus lutescens Persoon ex Fries

Cap membraneous, convex and depressed in the centre, with wavy and fringed margin; then expanded and infundibuliform, often with drooping margin; fibrillose and scaly; sooty-brown, sometimes tinged with orange. **Pseudo-gills** resembling wrinkles, decurrent, orange. **Stipe** like an upside-down cone; often irregular, compressed, furrowed; yellow or orange, sometimes tinged with green. **Flesh** tough, yellow. **Smell** strong. **Taste** mild. **Spores** white, 9–12 × 7–8 μ; elliptical. **Habitat** in the woods, in large colonies. **Edible/Poisonous** edible.

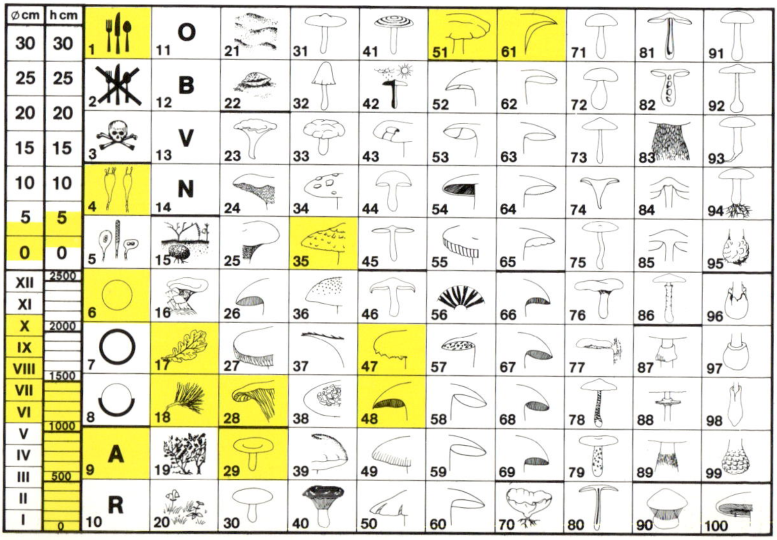

Carpophore globular, rough, dry; slightly pruinose, yellow; supported by a rough, stocky stipe, also yellow, gibbous. **Flesh** protected by a thick and tough peridium; firm, white, tinged with reddish, finally blackish and powdery. **Spores** brownish, 10–15 µ; globular, reticulated. **Habitat** on sandy soils by the seaside. **Edible/Poisonous** edibility unknown.

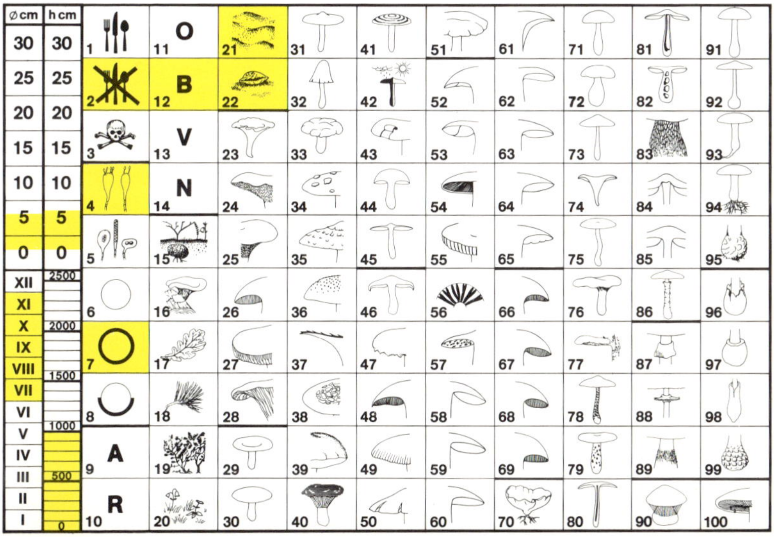

Parasol Mushroom
Cap globular, then expanded-umbonate; cuticle fibrillose, brownish, breaks off into scales disclosing the sub-cuticle which is whitish and fibrillose; fringed margin. **Gills** white then yellowish. **Stipe** white with brownish streaks and speckles, easily broken off from the cap; ring in various layers, becoming detached from the stipe and descending towards the foot. **Flesh** soft and white or tinged with pink in the cap; fibrous and darker in the stipe. **Smell** faint. **Taste** pleasant. **Spores** white, 14–18 × 18–12 μ; oval. **Habitat** in groups in grass at the edge of a copse, or in clearings in woods. **Edible/Poisonous** edible; of good quality when properly cooked; it is thought to be slightly poisonous when raw.

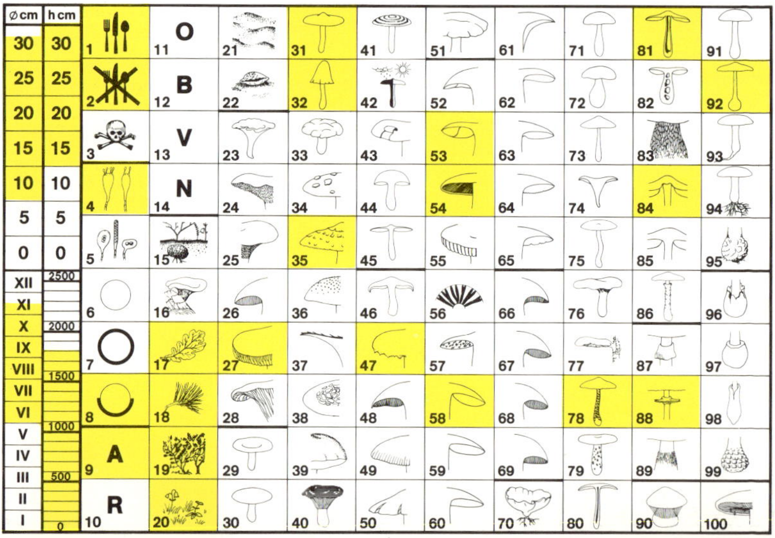

Ragged (or Shaggy) Parasol
Cap globular, then expanded; felt-like cuticle soon brownish, dissolving into rough scales laid out like roof tiles, uncovering the whitish and fibrillose sub-cuticle; fringed margin. **Gills** creamy, then reddish or rusty-brown. **Stipe** concolorous with cap, detachable, fibrillose; greyish, torn and free ring. **Flesh** tender and white in the cap, becomes brownish when exposed to the air. **Smell** sometimes slightly disgusting. **Taste** pleasant. **Spores** white, 11–13 × 6–8 μ; ovoid. **Habitat** mainly in sunny glades. **Edible/Poisonous** edible; of good quality when properly cooked; it is thought to be somewhat toxic when raw.

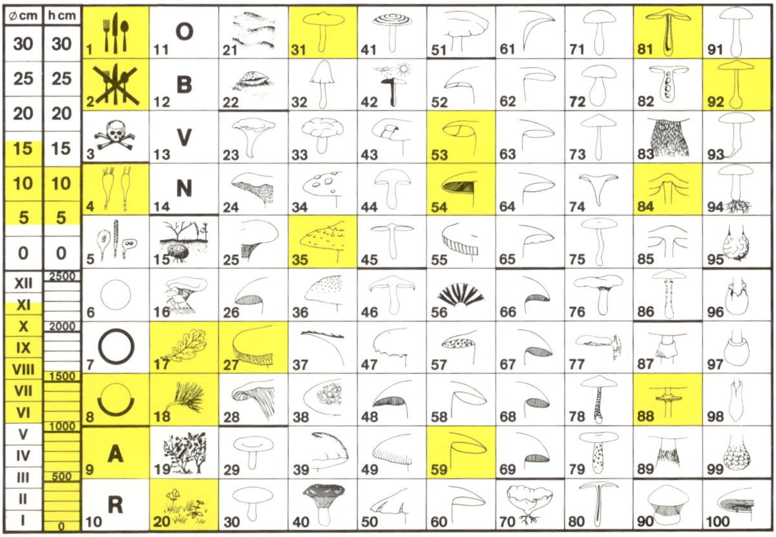

Syn. *Macrolepiota excoriata* (Schff. ex Fr.)
Cap globular, then expanded-umbonate; ash-grey to ochraceous cuticle, dissolving into floccose scales and exposing the paler sub-cuticle; fringed margin. **Gills** white then creamy. **Stipe** white then ochraceous, fibrillose, sometimes smooth; single ring, white then brownish, free. **Flesh** dry, milk-white. **Smell** pleasant. **Taste** pleasant. **Spores** white, 11–15 × 7·5–10 μ; ovoid. **Habitat** mainly in sunny glades and pastures. **Edible/Poisonous** edible; of good quality when cooked; probably slightly toxic when raw.

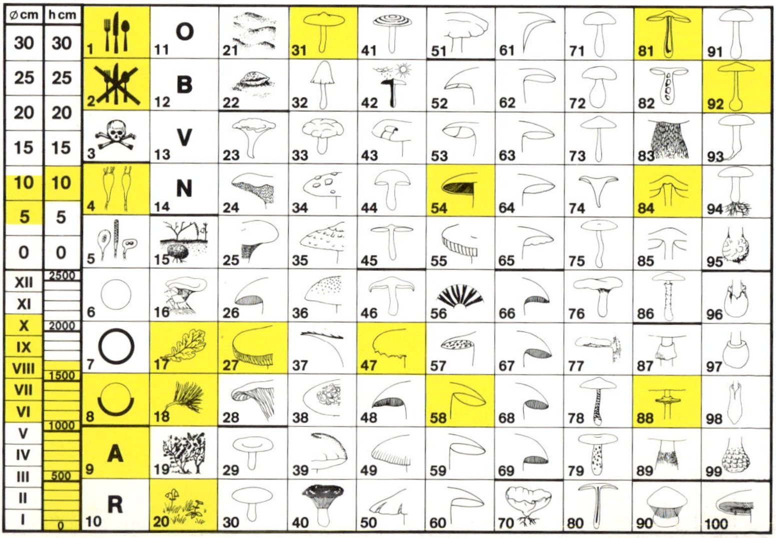

Lepiota helveola Bresadola

Cap from pinkish to brownish, dry, with scale on concentric rings. **Gills** white. **Stipe** concolorous with cap, with fugacious ring. **Flesh** white, becoming tinged with pink when touched or exposed to the air. **Smell** almost nil; when kept in a box for some time it reveals a sweet scent. **Taste** almost nil. **Spores** white, 8–10 × 5–6 μ; elliptical. **Habitat** in woods, but more frequently in parks and gardens. **Edible/Poisonous** poisonous; probably deadly.

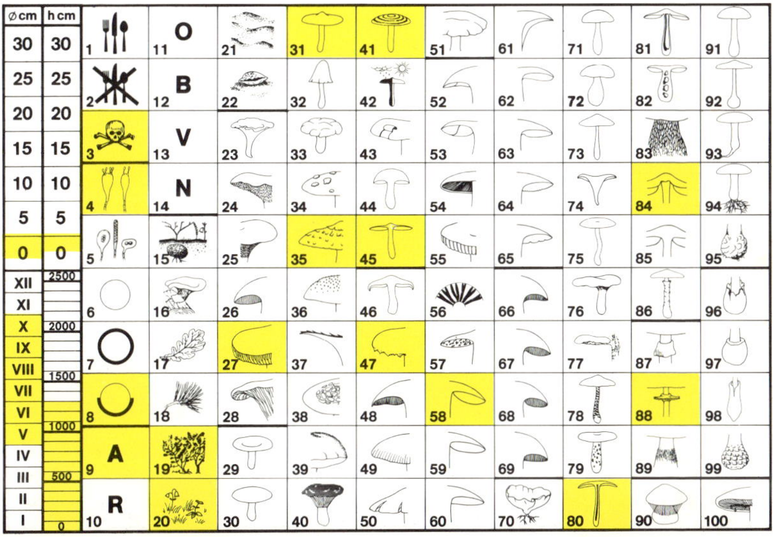

Limacella guttata Persoon ex Fries

Cap long, convex, rarely plane; usually with festooned margin; easy to peel; viscid in wet weather, otherwise dry; cream to nut-brown. **Gills** white tinged with cream. **Stipe** with bulbous foot; ringed; white; between the ring and the cap it exudes tiny droplets which turn into sooty spots; full. **Flesh** firm in the cap, fibrous in the stipe; white. **Smell** of flour. **Taste** of flour. **Spores** white, 5–6 × 4·5–5 μ; globular. **Habitat** under conifers from the end of June until the end of November; rarely under frondose trees. **Edible/Poisonous** edible.

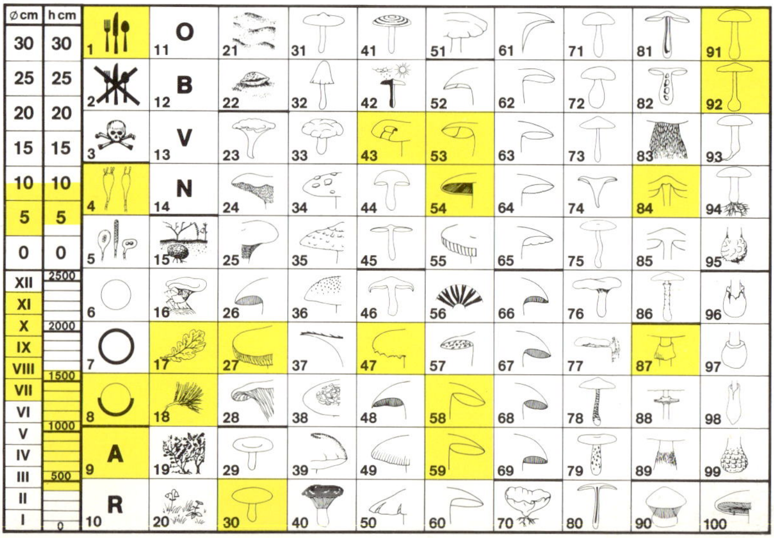

Cultivated Mushroom

Cap hemispherical, then expanded; white, becoming brownish if bruised or with age. **Gills** whitish, then reddish-pink and finally black. **Stipe** whitish, sometimes hollow at the foot. **Flesh** tender, white, becoming pinkish when exposed to the air. **Smell** of fruit. **Spores** brownish, 6–9 × 4–6·5 μ; elliptical. **Habitat** on soils rich in organic matter; cultivated on a large scale. **Edible/Poisonous** edible; of good quality.

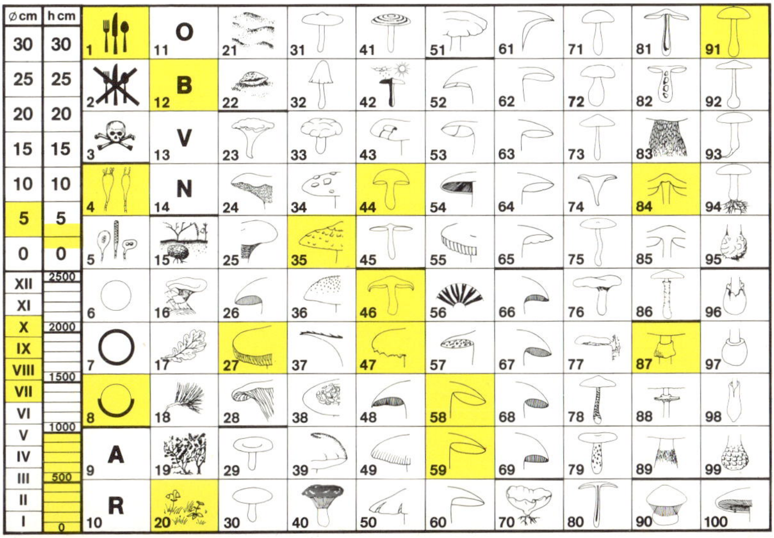

Cap bell-shaped, then expanded, often retaining a prominent apex; from whitish to brownish, the apex covered in ash-grey, shiny and fugacious powder; when old it shows deep radial furrows at the margin. **Gills** with dented edges; concolorous with cap. **Stipe** white, striate, with small fugacious ring, becoming concolorous with cap. **Flesh** yellowish, tender in the cap, fibrous in the stipe. **Smell** mild. **Taste** pleasant. **Spores** nut-brown or cinnamon, 10–13 × 7–9 μ; elliptical. **Habitat** mainly in the mountains under pines; sometimes on the plains under frondose trees. **Edible/Poisonous** edible; of good quality.

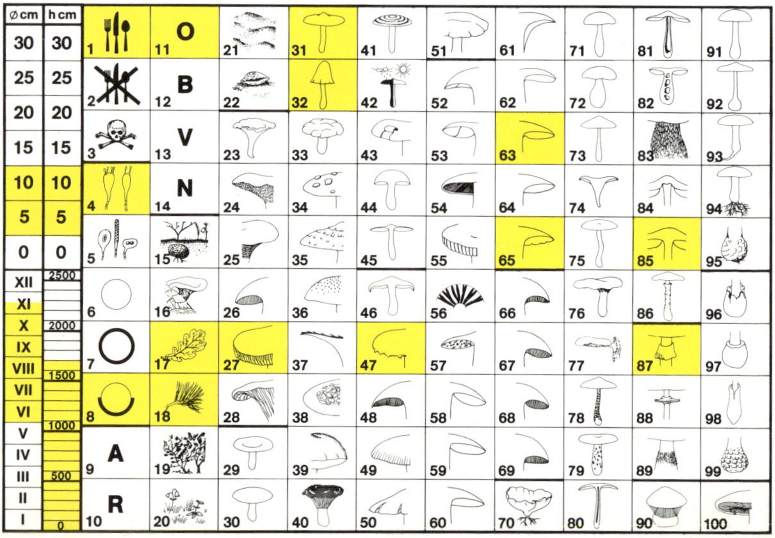

Agrocybe aegirita Briganti

Cap hemispherical, then plane or even slightly concave; from brown to whitish; in spring more often brownish, in autumn whitish; smooth and silky; with age or in dry weather it cracks from the middle. **Gills** whitish then brownish, with a lighter and irregular edge. **Stipe** arched, with the foot pressed against that of other individuals in the same tuft; white in the shade, otherwise more or less concolorous with the cap; with a wide, white and fugacious ring. **Flesh** white, then nut-brown; tender in the cap, fibrous in the stipe. **Smell** pleasant. **Taste** of nuts. **Spores** nut-brown, 9–11 × 6–7 μ; elliptical. **Habitat** from spring to late autumn on the trunks of frondose trees; in tufts. **Edible/Poisonous** edible; of good quality.

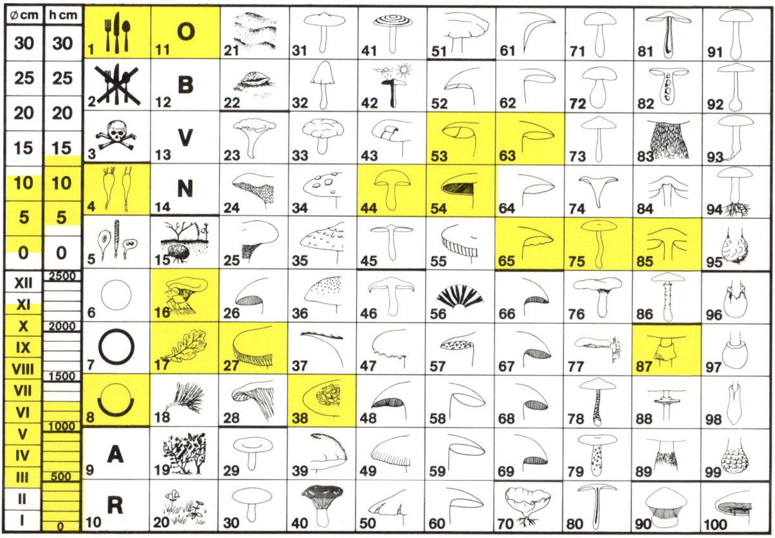

Early Agrocybe

Cap hemispherical, then plane or even slightly concave but still rather umbonate; often with fringed margin; from yellowish-ochraceous to brown in humid weather, paling to whitish in dry weather. **Gills** ash-white, then ash-grey-violaceous and finally a more or less dark nut-brown. **Stipe** slender, whitish, shiny, with brownish fibrils; with whitish ring or ring area. **Flesh** white or whitish, tender in the cap, fibrous in the stipe. **Smell** mild. **Taste** of flour, but faint. **Spores** light nut-brown, 8–11 × 5–7 μ; oval-oblong. **Habitat** from spring to midsummer in and out of woods; more frequent in rainy weather. **Edible/Poisonous** edible.

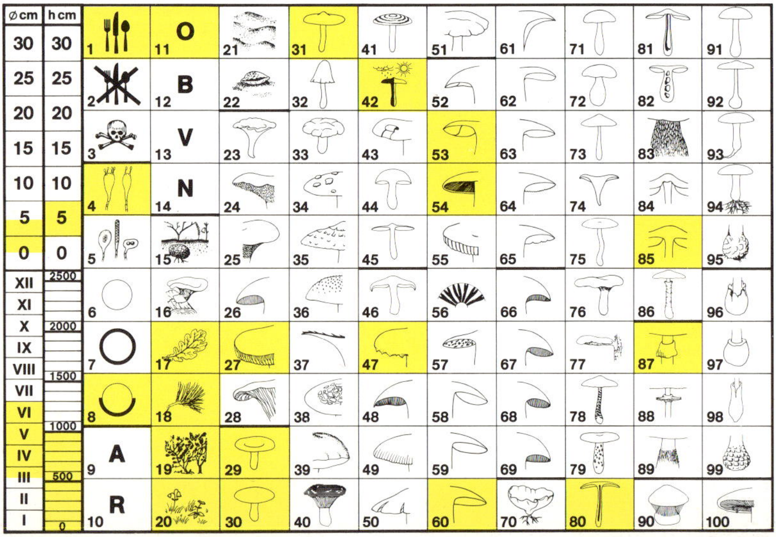

Fairy Cake Hebeloma

Cap plane to convex with dropping margin and the cuticle protruding beyond the gills; viscid in wet weather; yellowish or pink-brownish, darker in the centre. **Gills** white, tinged with nut-brown, then ochraceous; in wet weather they exude droplets which coagulate later into rusty spots; pruinose edge. **Stipe** more or less concolorous with the cap; pruinose or scaly at the top; full, then with marrow. **Flesh** flaccid and whitish. **Smell** of radishes. **Taste** of radishes, bitterish. **Spores** brownish, 11–12 × 5–6·5 μ; elliptical or amigdaliform. **Habitat** both in woods and in the open, in groups; common. **Edible/Poisonous** edible, though poisonous to some people.

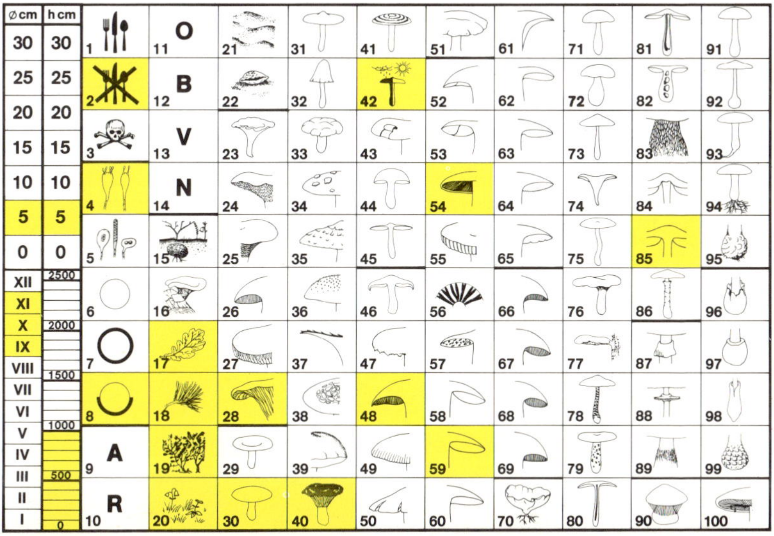

Rooting Shank

Syn. *Oudemansiella radicata* (Fr.) Sing. *Collybia radicata* (Fr.) Quél.

Cap umbonate-convex, albeit moderately, even when plane; thin margin; brown, sometimes shaded with olivaceous; viscid, easy to peel. **Gills** distant, white shaded with pink. **Stipe** tall, cylindrical, with a fusiform foot which descends deep into the soil like a root; white with reddish tinges; smooth and shiny. **Flesh** tender and fragile in the cap, fibrous and tough in the stipe. **Smell** imperceptible. **Taste** imperceptible. **Spores** white, 12–16 × 9–12 μ; oval. **Habitat** from spring to autumn both under frondose trees and not far from them, in meadows; alone or in small groups. **Edible/Poisonous** edible, but poor and not to be recommended.

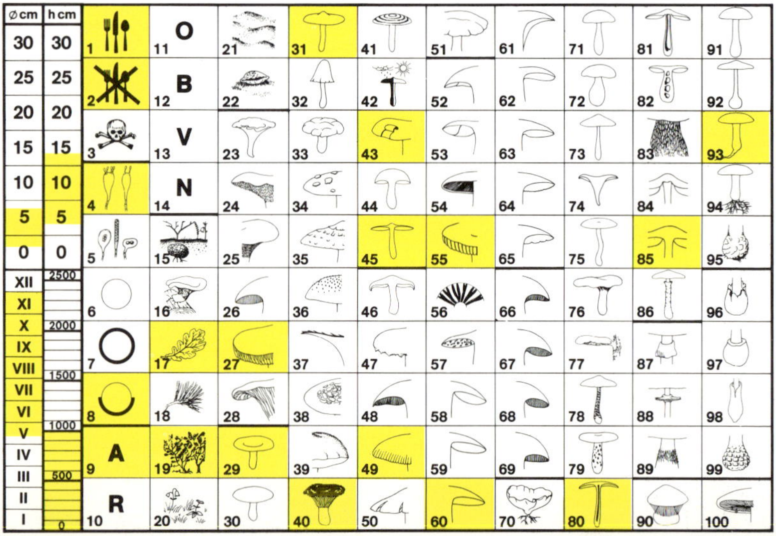

Fairy-ring Mushroom

Cap slightly umbonate even when it becomes moderately concave; glabrous, nut-brown, at times shaded with pink. **Gills** white or creamy. **Stipe** slender, lighter than the cap, fibrous and slightly twisted. **Flesh** rather tough in the cap, fibrous in the stipe; white or tinged with cream. **Smell** mild. **Taste** pleasant. **Spores** white, 7·5–9·5 × 5·5–6 μ; elliptical. **Habitat** in open spaces, among grass, both on low and high ground; in groups. **Edible/Poisonous** edible; of good quality once the fibrous stipes have been removed (can be confused with poisonous *Clitocybe* species).

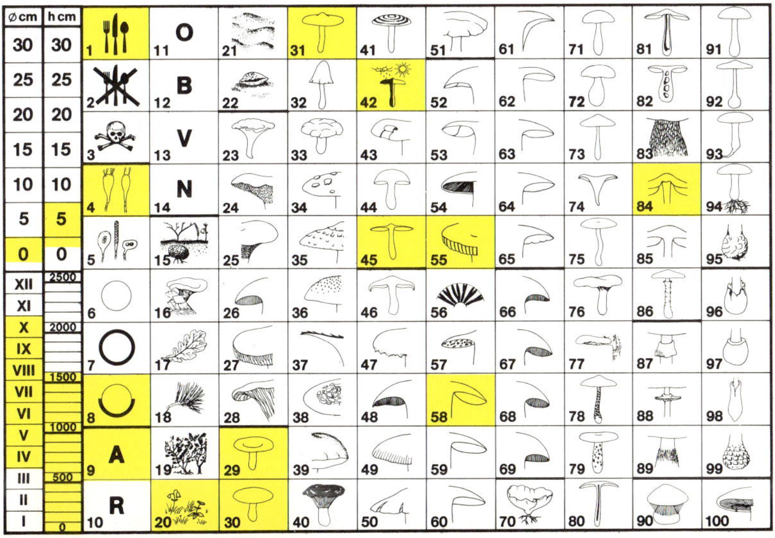

Melanoleuca grammopodia Bulliard ex Fries

Ring Agaric

Cap always umbonate even when it becomes slightly concave; from greyish-brown to pale ochraceous-grey. **Gills** very crowded, white. **Stipe** a little lighter than the cap, with darker fibrils running from top to bottom; full, then with fibrous marrow. **Flesh** rather firm in the cap, fibrous in the stipe; white. **Smell** slightly disgusting. **Taste** mild. **Spores** white, 8–11 × 5–6·5 μ; elliptical with apex. **Habitat** both on low and on high ground, woods or meadows; in circles or semi-circles; common. **Edible/Poisonous** only the cap is edible, but poor even when properly cooked.

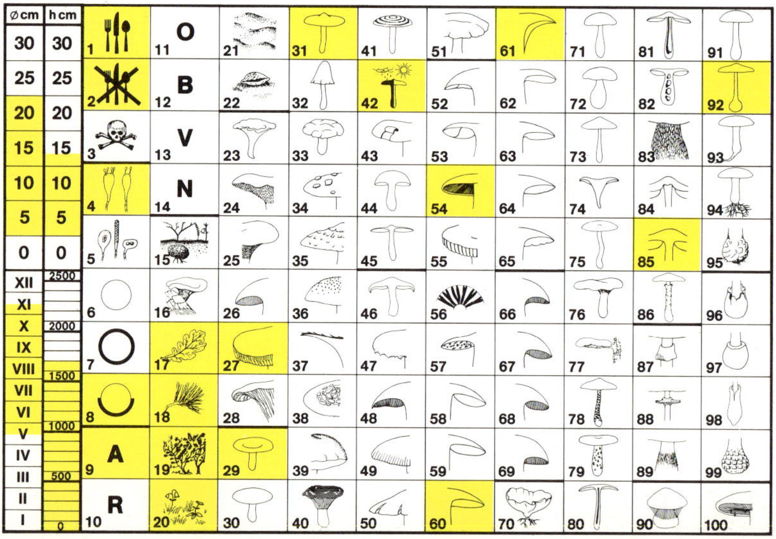

Common Funnel Cap

Syn. *Clitocybe gibba* (Pers. ex Fr.) Kumm

Cap slightly umbonate even when deeply concave, infundibuliform; fibrillose, dry, ochraceous. **Gills** white, then creamy. **Stipe** at times swollen at the foot, where it is also covered in floccules and mycelial cords. **Flesh** somewhat compact in the cap, fibrous in the stipe; white. **Smell** pleasant. **Taste** pleasant. **Spores** white, 4·5–7·5 × 3–5 μ; ovoid with apex. **Habitat** woods, heaths and grass in groups until early winter. **Edible/Poisonous** edible; of good quality.

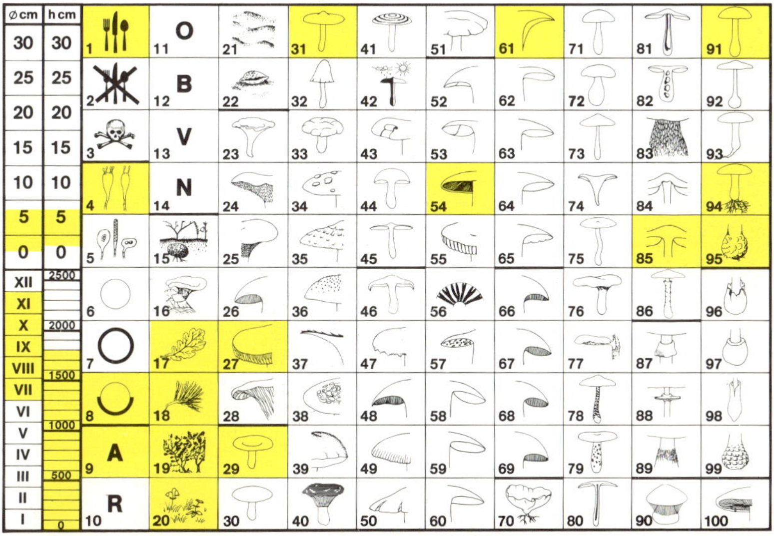

Cap moderately convex or plane; irregular, slightly striate, with involute then drooping margin; viscid cuticle, later slightly rough; fibrous, with some scales; ochraceous, with a paler margin. **Gills** white, with ash-grey shades, waxy; then creamy. **Stipe** white, pruinose or floccose in the upper part; yellowish in the lower half and soiled with vegetal particles; full. **Flesh** rather fragile in the cap; almost woody in the foot; white with gelatinous streaks which become yellowish when exposed to the air. **Smell** of flour and radishes. **Taste** somewhat bitter. **Spores** white, 4·5–6 × 3·5–4·5 μ; ovoid or amigdaliform. **Habitat** under frondose trees. **Edible/Poisonous** edible, even when raw.

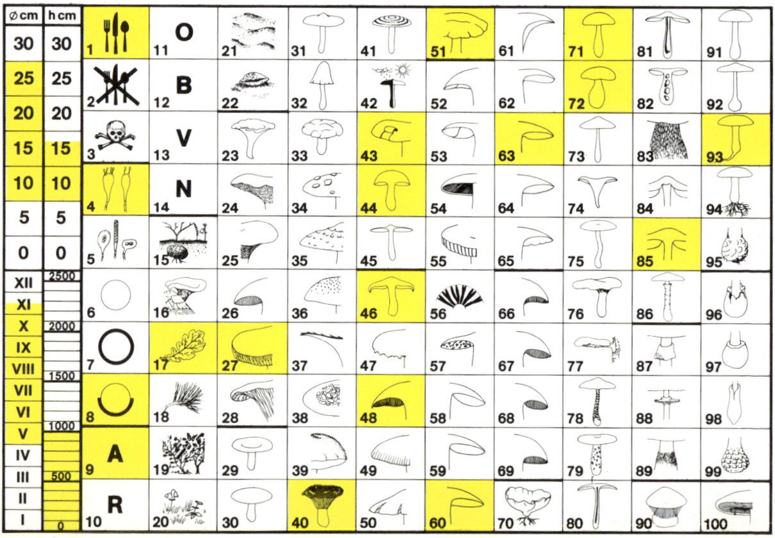

Cap finally depressed or even cup-shaped, with lobate margin; yellowish-pink, fuscous or ochraceous; covered in rusty scales, later pruinose. **Gills** at first barely visible, then clear and thin; forked; white then creamy. **Stipe** club- or spindle-shaped, ending like a truncated root; sometimes eccentric; cream or brownish, darker at the foot; at times scaly like the cap. **Flesh** white and firm in the cap, whitish and tough in the stipe. **Smell** pleasant. **Taste** pleasant. **Spores** white, 10–14 × 4–5 μ; cylindrical or kidney-shaped. **Habitat** on the stumps of various plants; rare, but recurring in its stations. **Edible/Poisonous** edible only when very young.

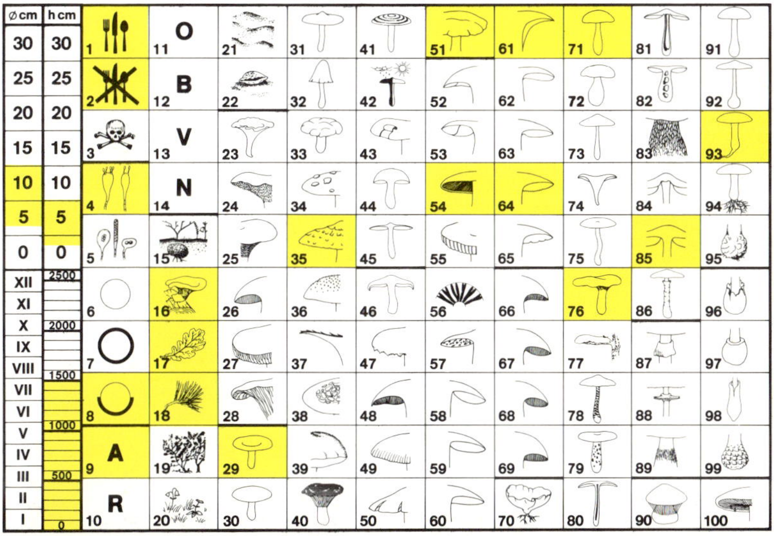

Sooty Milk Cap

Cap with involute and then drooping margin, even when it has become infundibuliform; often gibbous; glabrous, but when seen through the lens it appears velvety; dry, ochraceous and and brownish, becoming lighter with age. **Gills** sparse and thin; ivory-white then reddish-creamy. **Stipe** with mycelial cords at the foot; more or less concolorous with the cap; full then stringy. **Flesh** creamy-white, turning yellowish on exposure to the air. **Smell** faint; of fruit. **Taste** slightly bitter. **Milk** copius, white, becoming reddish in the air; acrid. **Spores** yellowish, 8–9·5 × 8–9 μ; globular, reticulated. **Habitat** in the woods, in small groups. **Edible/Poisonous** inedible.

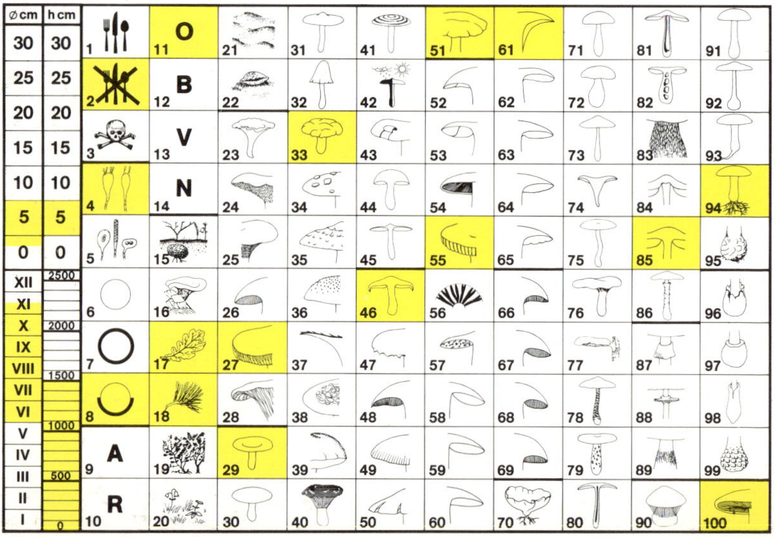

Bitter Bolete or Sponge Cap
Syn. *Tylopilus felleus* (Bull. ex Fr.) Karst.
Cap thick and convex, sometimes depressed; dry, velvety, ochraceous. **Tubes** 1·5–3cm long; initially tangent to the stipe, then free; white, then tinged with reddish. **Pores** round, then angular; rather large; white, then reddish and finally rusty. **Stipe** with swollen and even bulbous foot; somewhat irregular; concolorous with the cap or slightly paler; covered in brownish lattice of oblong, angular and protruding meshes; floccose foot; full. **Flesh** soon flaccid, white. **Smell** faint. **Taste** all the more bitter as the mushroom grows old. **Spores** pinkish, 10–16 × 4–5 μ; elliptical. **Habitat** from late spring to late autumn under frondose trees, even in the hollows of some trunks; common, but rare under conifers. **Edible/Poisonous** inedible.

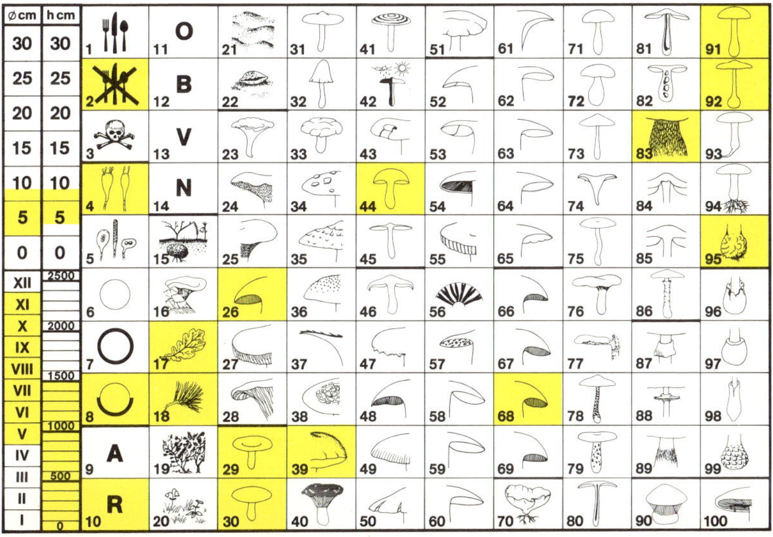

Dryad's Saddle or Scaly Polyporus

Cap like a semi-circular shelf which can reach a diameter of 50cm; ochraceous, more or less faded; with brownish scales laid on concentric rings. **Tubes** short, 0·1–0·4cm, decurrent; white then creamy. **Pores** small, then becoming larger and angular; white, then yellowish. **Stipe** usually lateral, stocky, dark or covered in black velvet, but yellowish at the top; full. **Flesh** firm, slightly watery, then elastic and tough; whitish. **Smell** pleasant. **Taste** mild, sour-sweet. **Spores** white, 11–15 × 4–5·5 μ; elliptical. **Habitat** on the stumps of various frondose trees in the spring, but can appear again in the autumn; rare on stumps of conifers; occasionally in tufts. **Edible/Poisonous** edible when very young; not edible otherwise.

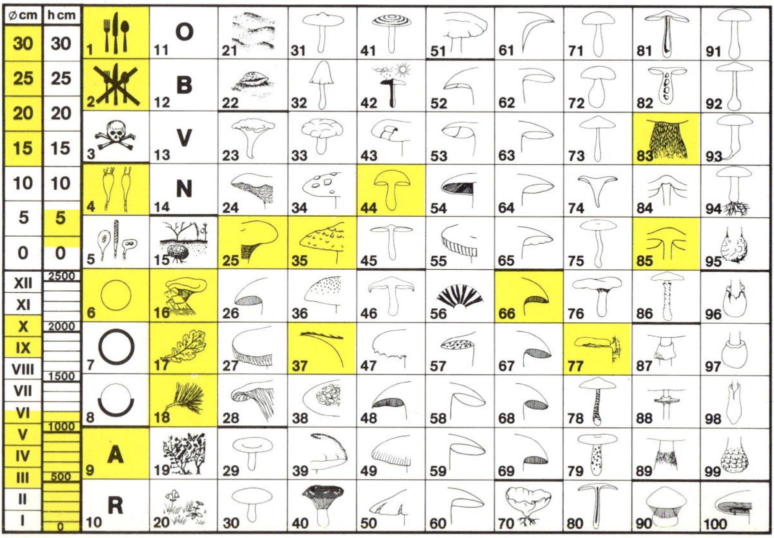

It consists of a thick trunk, whitish and pruinose, from which grow several oblique branches, which are short, flat and white; on the tips of the branches grow horizontal caps like irregular spatulae with wavy margins; on the upper surface these caps are grey or brownish, velvety, slightly wrinkled, with dark fibrils; underneath they are lined with short tubes, each ending in a pore, small, round and white, and later oblong and grey **Flesh** fragile and fibrous in the cap; elastic and tough in the branches; white. **Smell** aromatic, sharp. **Taste** pleasant. **Spores** white, 5–6·5 × 4–5 μ; oval. **Habitat** on the stumps of various frondose trees in late summer and autumn. **Edible/Poisonous** edible; of good quality while still young; not edible when ageing.

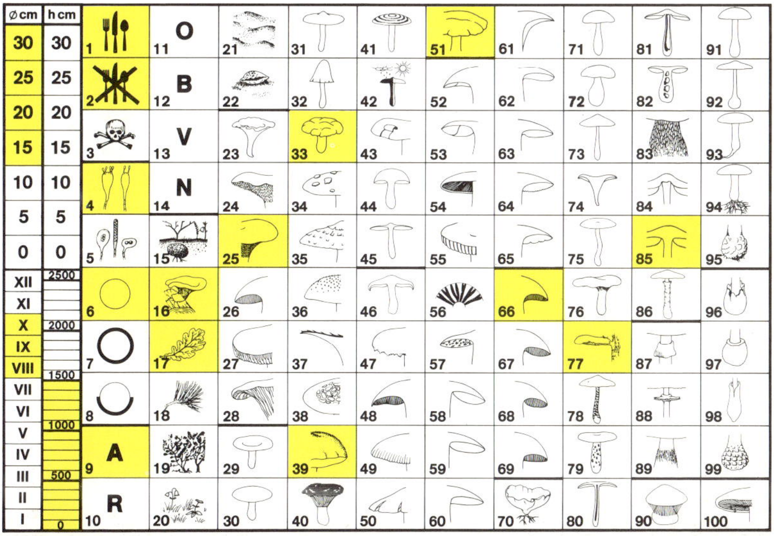

Wood Hedgehog
Syn. *Hydnum repandum* L. ex Fr.
Cap considerably irregular, pruinose or velvety; yellowish-pink or orange but at times whitish. **Spines** fragile, decurrent, detachable in groups; yellowish-white. **Stipe** often irregular, sometimes eccentric; concolorous with the cap; full. **Flesh** fragile, whitish or yellowish. **Smell** pleasant. **Taste** slightly bitter. **Spores** white tinged with cream, 7–9 μ; globular. **Habitat** in the woods, in groups; common. **Edible/Poisonous** edible; of good quality while still young.

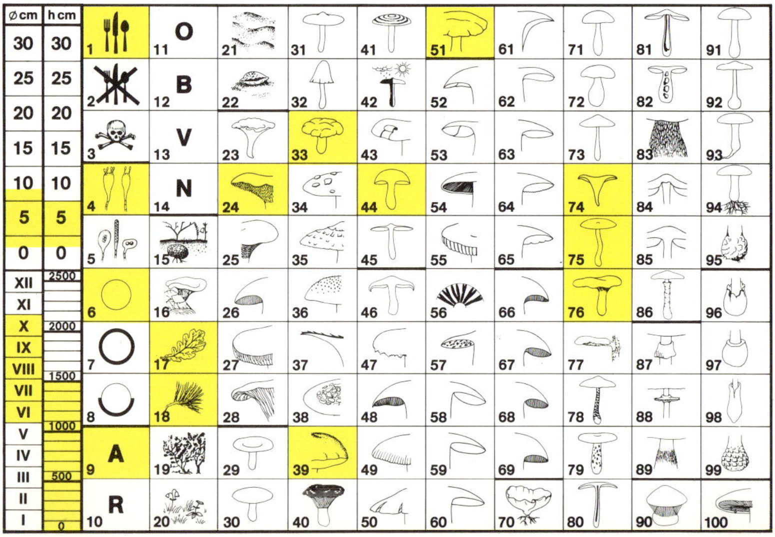

Panther Cap or False Blusher

Cap globular then expanded, from brown to ochraceous, viscid in moist weather; becomes drier with age, loses the warts and fades. **Gills** white. **Stipe** white, soon becomes tubular, with wide, soft ring, double-rimmed, fugacious and white; bulbous foot covered by volva, with a neat brim; above the foot, floccose remnants in a circle or semi-circle, more or less oblique. **Flesh** white and fragile. **Smell** mild, but disgusting in old specimens. **Taste** mild, but acrid when old. **Spores** white, 9–12 × 7–8 μ; elliptical. **Habitat** not very common under conifers; rare under frondose trees. **Edible/Poisonous** poisonous; can be deadly.

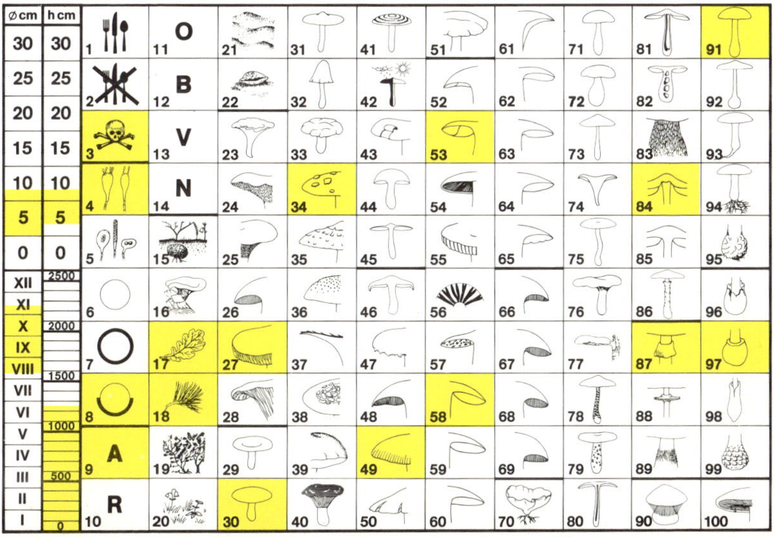

The Blusher

Cap globular, gradually becoming less convex; pinkish, sometimes from whitish to brownish; with a striate margin later. **Gills** white, becoming rusty with age. **Stipe** white tinged with pink below the ring; white ring, often with brownish edge, striate; bulbous foot showing volva remnants, scaly and whitish. **Flesh** white, turning reddish in the air. **Smell** almost imperceptible. **Taste** occasionally slightly acrid. **Spores** white, 7·5–11 × 6–7·5 μ; elliptical-oblong. **Habitat** in the woods. **Edible/Poisonous** edible only when cooked; slightly toxic when raw.

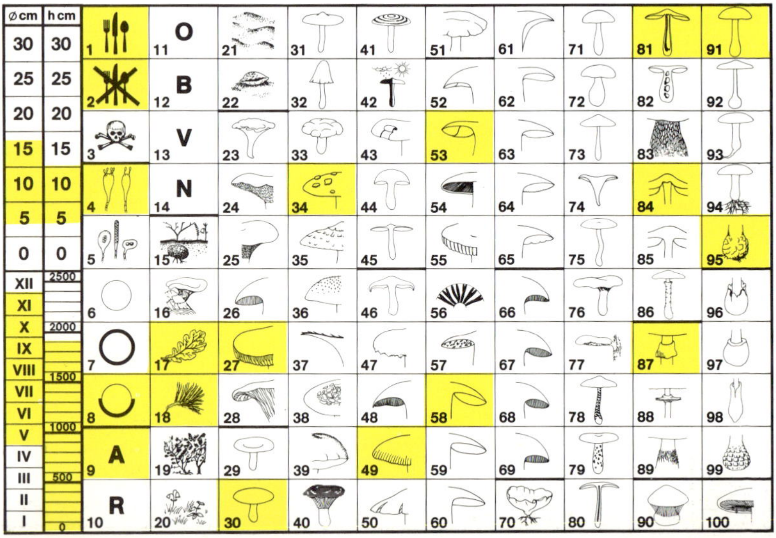

Cap brownish, scaly, with an often festooned margin. **Gills** pure white. **Stipe** brownish, but white from the ring upwards; thin and fugacious ring. **Flesh** white. **Smell** a bit unpleasant. **Taste** a bit acid. **Spores** white, 5–7 × 2·5–3·5 μ; cylindrical. **Habitat** in soils fertilised with organic matter. **Edible/Poisonous** inedible because of its disgusting taste.

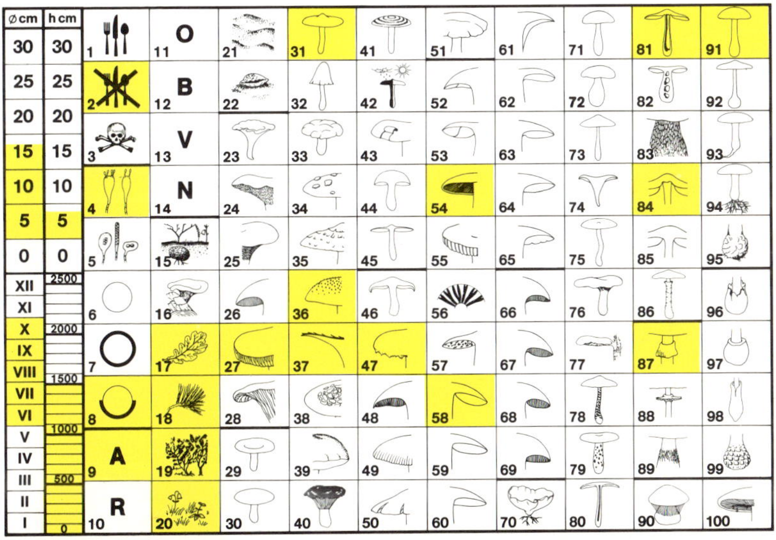

Cortinarius orellanus Fries

Cap convex, finally plane, remaining slightly protruberant at the apex and occasionally slightly wavy; brown tending to reddish or orange; fibrillose, sometimes scaly in the centre. **Gills** deep, thick, distant, rusty. **Stipe** a little lighter than the cap, fibrillose; thin, yellow cortina soon disappearing. **Flesh** yellowish, tinged with red; paler in dry weather. **Smell** of turnips. **Taste** slightly acid after lengthy chewing (do not swallow!). **Spores** nut-brown, 10–12 × 5–6·5 μ; elliptical. **Habitat** in all woods; luckily rare. **Edible/Poisonous** deadly.

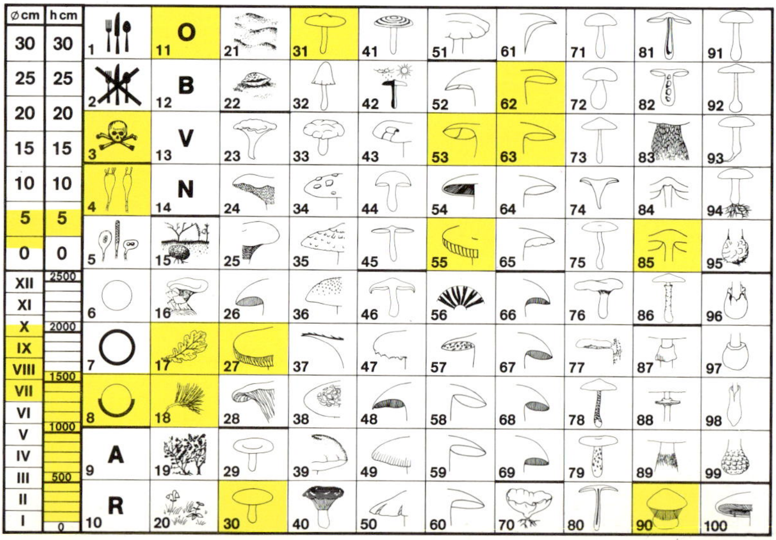

Roman Shield Entoloma

Cap with involute margin; then expanded, but umbonate with wavy margin; from grey to sooty in wet weather, fading in dry conditions to whitish; silky. **Gills** distant, with slightly wavy edge; white then pinkish. **Stipe** white, silky, soon fistulous. **Flesh** fragile in the cap, fibrous in the stipe; white then ash-grey. **Smell** pleasant; of flour. **Taste** pleasant; of flour. **Spores** pinkish, 8·5–11 × 7·5–10 μ; polygonal. **Habitat** from May to July, rarely later, usually in the grass in the shade of *Rosaceae*; in groups or small tufts. **Edible/Poisonous** edible; of good quality.

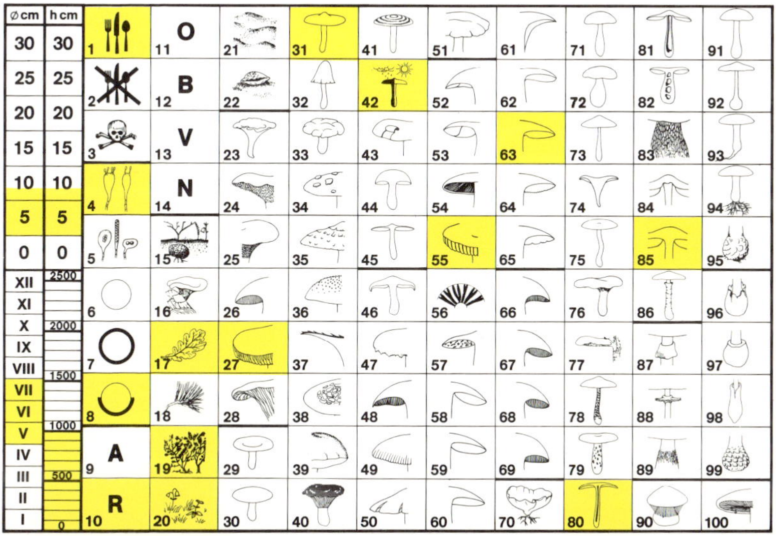

Tricholoma caligatum Viviani

Cap hemispherical, fibrillose, brown, with involute margin; then expanded but with prominent apex and festooned margin; in time the cuticle becomes scaly and the white sub-cuticle appears between the brownish scales. **Gills** white, then creamy with some spots on the edge. **Stipe** covered in brownish scales which end at the top in a rudimentary, fugacious ring leaving the white surface. **Flesh** firm in the cap, fibrous in the stipe; white but creamy under the cuticle. **Smell** of fruit; when dry it smells delicious. **Taste** slightly bitter. **Spores** white, 6–8 × 4–5·5 μ; elliptical. **Habitat** heathland. **Edible/Poisonous** edible.

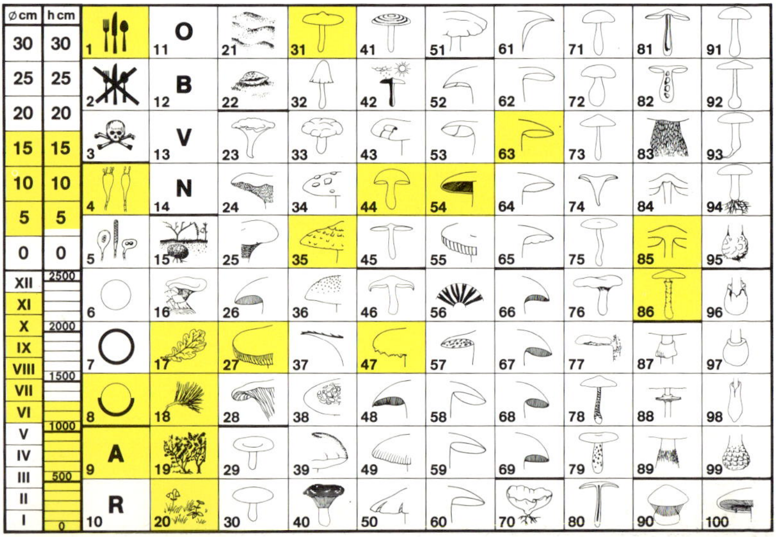

Sandy Tricholoma

Cap often irregular, with wavy margin; viscid in wet weather; various shades of brown with darker patches; easy to peel. **Gills** white, detachable in blocks. **Stipe** white then brownish. **Flesh** firm in the cap, fibrous in the stipe; white, slightly brownish under the cuticle. **Smell** of flour. **Taste** a little bitter. **Spores** white, 5·5–6 × 3·5–4 μ; elliptical. **Habitat** along the roots of poplars, on sandy soil, in circles. **Edible/Poisonous** inedible fresh, owing to its bitterness; edible and of good quality when preserved in oil or vinegar.

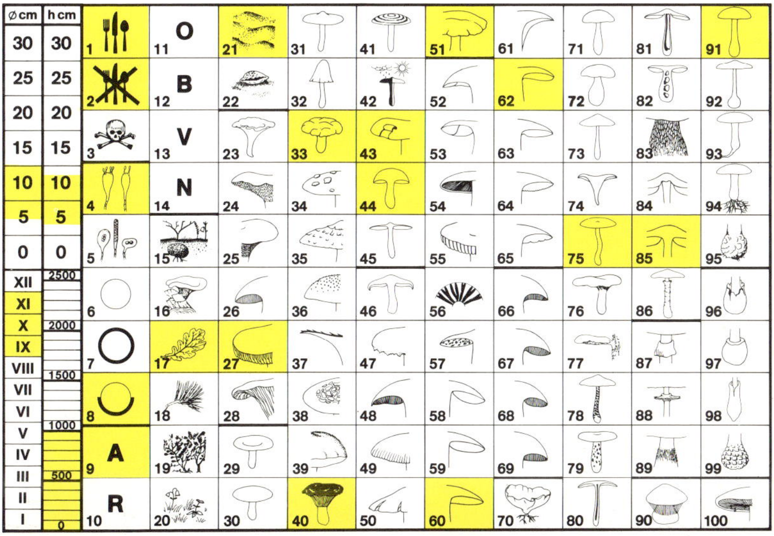

Soap Tricholoma

Cap usually with wavy margin, fissured in dry weather; green, grey, yellow or brownish. **Gills** white, then greenish; finally patched with brown. **Stipe** irregular, smooth, sometimes fibrillose or scaly; concolorous with the cap or slightly paler; full. **Flesh** firm, white, becoming reddish when exposed to the air, particularly in the foot. **Smell** of soap. **Taste** mildly of flour, or bitterish. **Spores** white, $5–6 \times 3{\cdot}5–4\ \mu$; oval. **Habitat** in the woods; in groups. **Edible/Poisonous** not to be recommended, although not poisonous, as it is rather disgusting even when properly cooked.

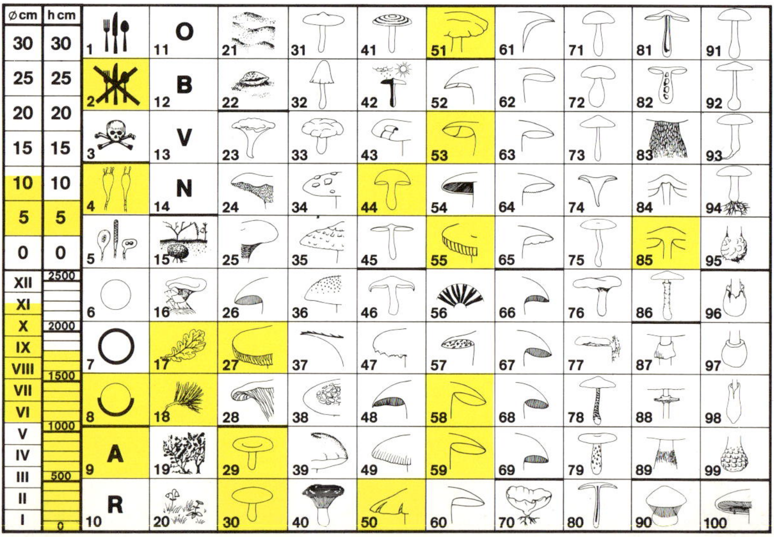

Armillaria tabescens Scopoli ex Fries

Cap umbonate, with lobate margin; ochraceous or brownish, covered with darker scales in the centre. **Gills** whitish, then pinkish-brown, finally brownish with white spots. **Stipe** fibrillose, concolorous with the cap. **Flesh** firm and white in the cap; fibrous and brownish in the stipe. **Smell** pleasant. **Taste** sweetish then bitterish. **Spores** white, 8–10 × 5–6 μ; ovoid. **Habitat** until early winter on stumps and on the ground; in tufts. **Edible/Poisonous** edible when cooked; of good quality.

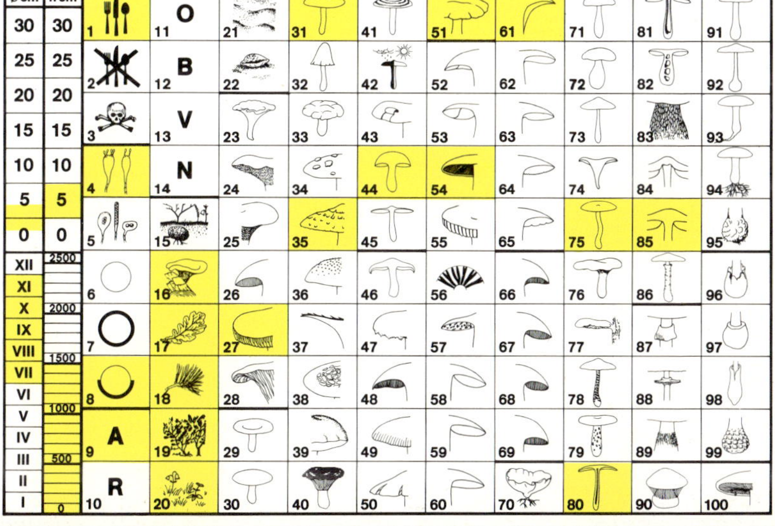

Clitocybe olearia (de Candolle ex Fries) Maire

Cap convex, with slight central depression and involute margin; then expanded and even concave or infundibuliform; smooth, shiny, pruinose; orange or fuscous, sometimes fibrillose. **Gills** yellow or orange, covered in yellowish-white powder by the ripe spores, phosphorescent. **Stipe** tapering towards the cap, often with pointed foot, pressed against the feet of the other individuals in the same tuft; a little lighter than the cap. **Smell** pleasant. **Taste** mild. **Spores** yellowish, 5–7 × 4·5–5·5 μ; oval or almost globular. **Habitat** on the stumps of frondose trees; in tufts. **Edible/Poisonous** poisonous.

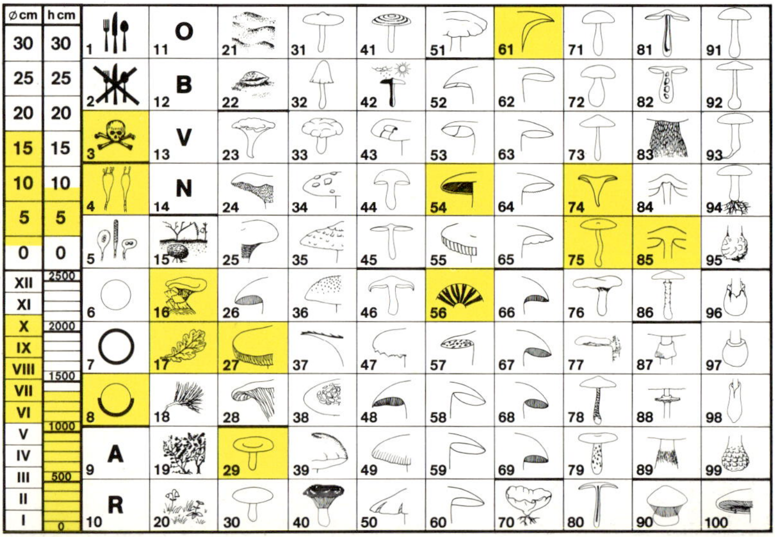

Cap convex, then plane and finally even depressed; fringed margin: dry, fibrillose or scaly, often with whitish, radial lines, clearer towards the margin; brown or fuscous. **Gills** white then yellow. **Stipe** often slightly eccentric and curved; scaly, whitish or brownish, with cortina remnants. **Flesh** white, slightly brown under the cuticle; firm in the cap, fibrous and tough in the stipe. **Smell** pleasant. **Taste** slightly acrid. **Spores** white, 6–6·5 × 3–3·5 μ; elliptical-oblong. **Habitat** on the stumps of various frondose trees from spring to autumn. **Edible/Poisonous** edible; of good quality; cultivated in Japan.

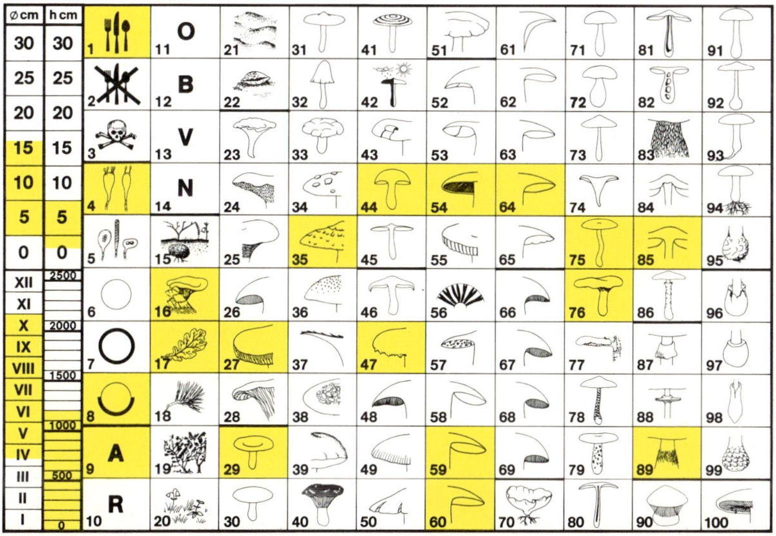

Rufous Milk Cap

Cap soon becomes plane, umbonate with involute margin; brick-red, occasionally velvety. **Gills** yellowish, then ochraceous and rusty. **Stipe** lighter than the cap. **Flesh** whitish in the cap, reddish in the foot. **Smell** faint. **Taste** strong and acrid. **Milk** white, scarce, acrid. **Spores** white tinged with yellowish, 8–9 × 6–7·5 μ; oval. **Habitat** under conifers, particularly in the mountains. **Edible/Poisonous** inedible.

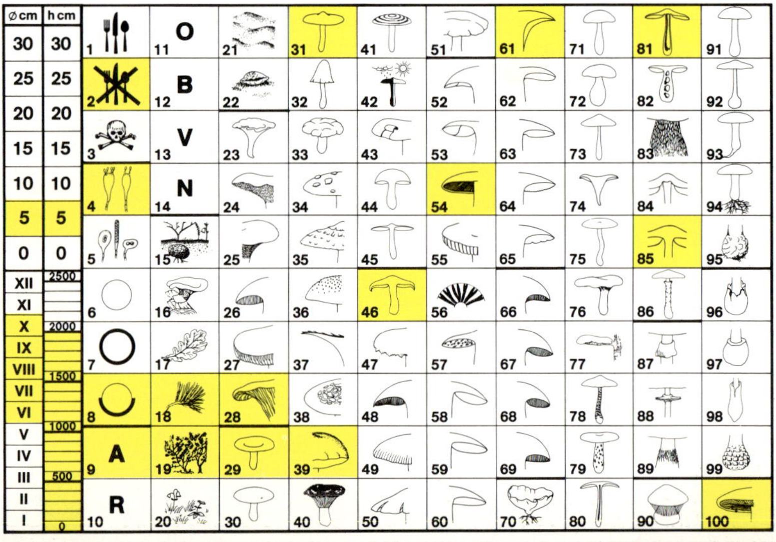

Cap convex, then plane or depressed or wavy; involute margin; glabrous, producing polygonal cracks in dry weather; yellow or orange with brown patches. **Gills** ivory-white, then creamy, becoming brownish when bruised. **Stipe** slightly lighter than the cap, with brown patches. **Flesh** whitish, becoming fuscous when exposed to the air. **Smell** salty. **Taste** salty. **Milk** abundant, white, sweetish. **Spores** white, 8–12 × 7–11 μ; globular, crested. **Habitat** in the woods, in groups; common. **Edible/Poisonous** edible.

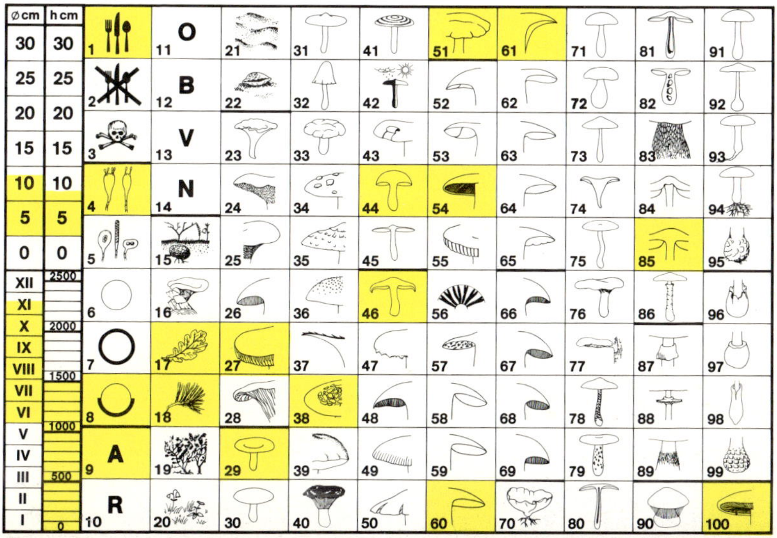

Bare-toothed Russula
Cap finally depressed, often irregular, with striate margin showing the tips of the gills; easy to peel; viscid in wet weather, otherwise dry and matt: from reddish-brown to violaceous-pink with yellowish patches; becoming paler with age. **Gills** joined to one another by small veins; white then creamy; the edge occasionally marked by rusty spots. **Stipe** finely wrinkled; white, becoming greyish where bruised; full. **Flesh** fragile, white. **Smell** of fruit. **Taste** of fruit. **Spores** white, 6–8 × 5–6·5 μ; elliptical. **Habitat** from May to November in the woods; in groups, seldom solitary; fairly common. **Edible/Poisonous** edible; of good quality.

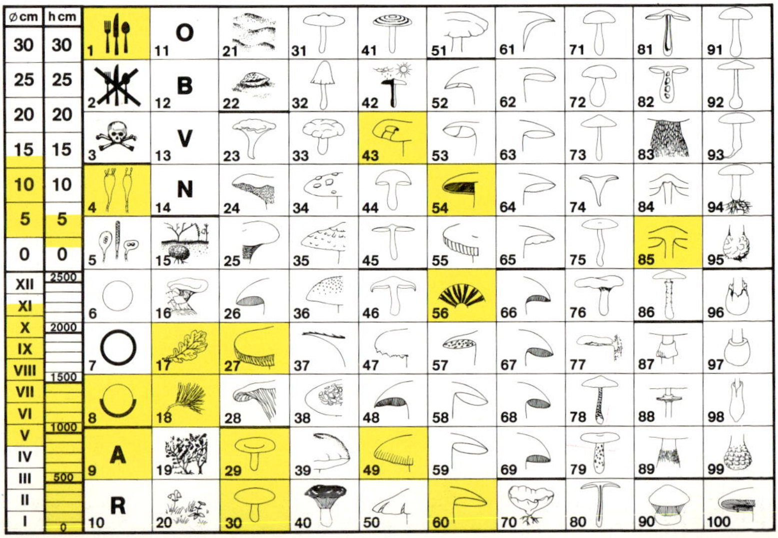

Cap eventually plane or even depressed, with striate margin; slightly viscid in wet weather, otherwise dry and matt; red or reddish-violet. **Gills** joined in places by small veins; whitish, then ochraceous. **Stipe** white or whitish; a bit rough; full, then tough. **Flesh** fragile, white. **Smell** faint. **Taste** pleasant. **Spores** tinged with golden yellow, 8–9 × 7–8 μ; globular. **Habitat** in summer and up to early winter in the woods; in groups. **Edible/Poisonous** edible; of good quality.

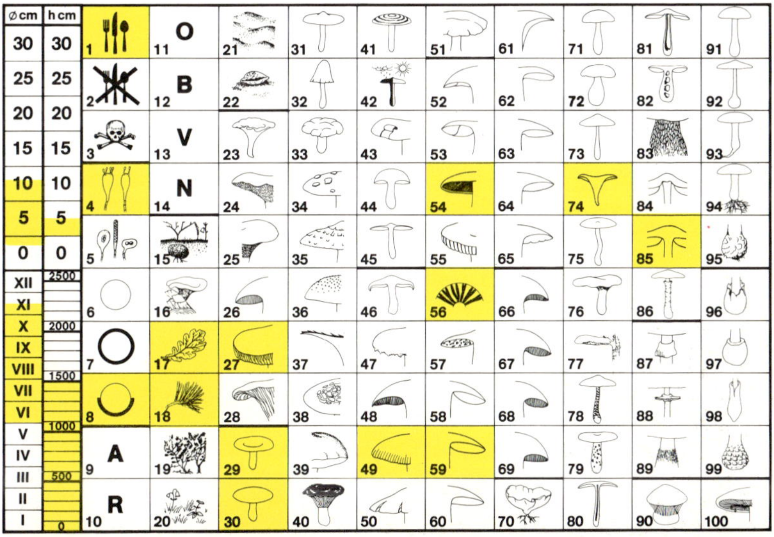

Cap campanulate, then expanded to umbonate, often with lobate and split margin; fibrillose, damp and almost viscous surface; grey or blackish, occasionally tinged with brown. **Gills** distant, connected by veins; white, with sooty edge. **Stipe** often curved, fibrillose, greyish or sooty. **Flesh** compact, white. **Smell** mild. **Taste** sweetish. **Spores** white, 8–9 × 5–6 µ; elliptical with apex. **Habitat** up to early winter under conifers; scattered. **Edible/Poisonous** edible.

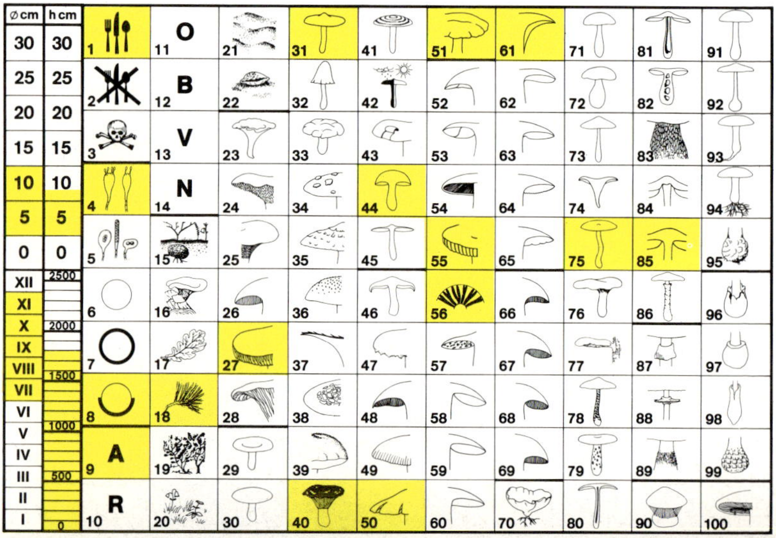

Cap umbonate with drooping margin; viscid, reddish-brown or violaceous-brown, with a few black spots. **Gills** distant, thick, deeply decurrent; greyish-yellowish, then violaceous-brownish. **Stipe** tapering towards the cap; fibrillose, with glutinous cortina, then with pseudo-ring left by the cortina; whitish but eventually blackish-brown; full. **Flesh** whitish, but bright yellow in the foot. **Smell** faint. **Taste** mild. **Spores** creamy, 17–22 × 6–8 μ; fusoid. **Habitat** under conifers, in groups. **Edible/Poisonous** edible; but poor.

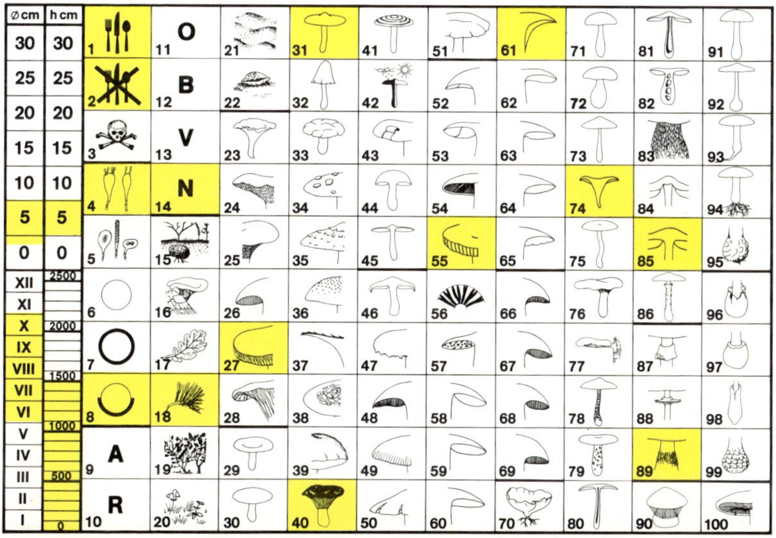

Inrolled Paxil or Brown Roll-rim

Cap eventually almost infundibuliform but with margin still involute or at least drooping; viscid, particularly in wet weather; from ochraceous to brown, blackish where bruised. **Gills** detachable in small groups (with age they detach themselves from the stipe); yellowish, brownish where bruised. **Stipe** tapering towards the cap; fibrillose, slightly lighter than the cap; full. **Flesh** soon flaccid; ochraceous, blackish-brown when exposed to the air. **Smell** faint. **Taste** bitterish. **Spores** rusty, 8–10 × 5–6 μ; oval. **Habitat** deciduous woodland, under birch on heaths; late summer to late autumn. **Edible/Poisonous** poisonous if raw or cooked when too old; edible as long as it is properly cooked when very young; can be preserved in oil or vinegar.

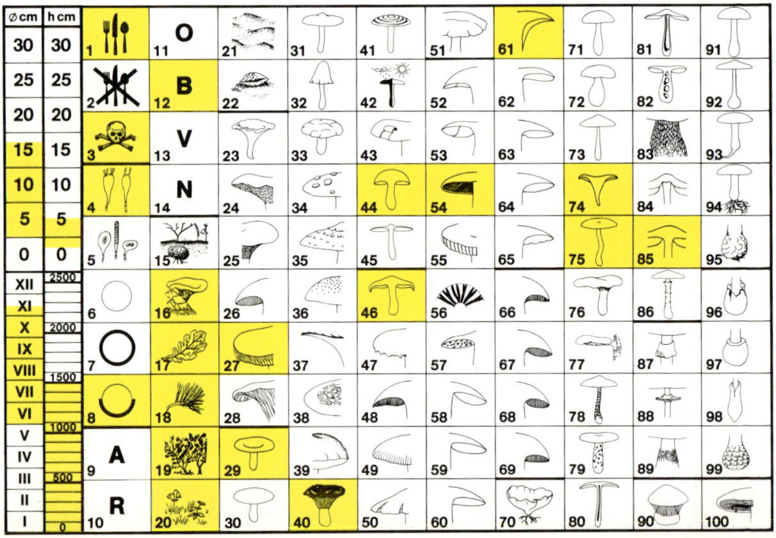

Black Velvet Paxil

Cap eventually slightly depressed but with margin still involute or drooping; dry, brown, velvety; later glabrous and faded; blackish when bruised. **Gills** yellow-olivaceous, then ochraceous. **Stipe** stumpy with rounded foot, often eccentric; concolorous with the cap, covered in blackish velvet; full. **Flesh** soon flaccid; yellowish-white. **Smell** of wood. **Taste** bitterish. **Spores** pale ochre, 5–6·5 × 3–4 μ; elliptical-globular. **Habitat** on the stumps of dead conifers, occasionally in small groups; remains fairly common. **Edible/Poisonous** not to be recommended, although not poisonous, as its taste remains unpleasant even when properly cooked.

Chestnut Bolete

Syn. *Gyroporus castaneus* (Bull. ex Fr.) Quélet

Cap convex, finally concave, with fissured margin; dry, velvety, chestnut-brown or orangey-brown. **Tubes** short and white, then yellowish. **Pores** tiny, round, white and later yellowish, eventually stained with ochre. **Stipe** often irregular, glabrous or slightly velvety; concolorous with the cap; with marrow then hollow, eventually becoming a hard and fragile bark. **Flesh** white, later creamy. **Smell** faint. **Taste** of hazelnuts. **Spores** whitish, 8–11 × 4·5–6·5 μ; elliptical. **Habitat** in the woods, in groups or alone. **Edible/Poisonous** edible; of good quality as long as not too old.

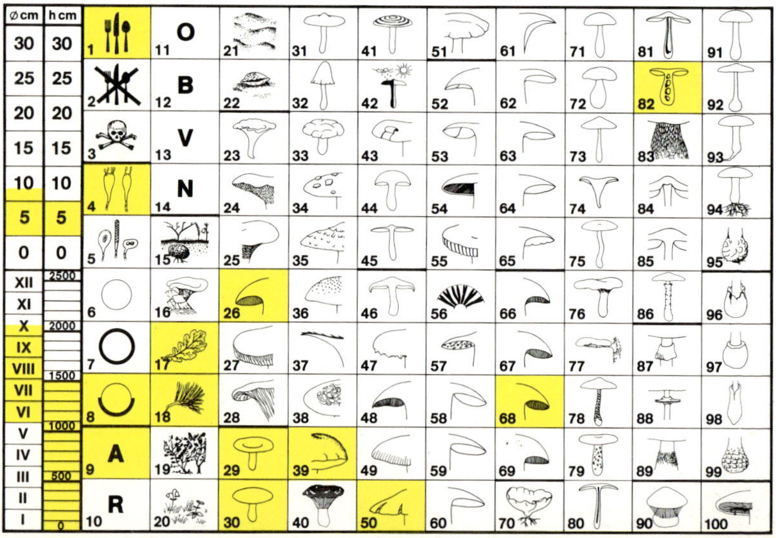

Cep or Penny-bun

Cap hemispherical, from whitish to ochraceous. **Tubes** 1·5–3cm long; soon free; thin, detachable in small groups; white, then yellowish and eventually olivaceous. **Stipe** pear-shaped or cylindrical, slightly paler than the cap; with white reticulations, very fine in the lower part; full. **Flesh** firm, later flaccid; white, discoloured near cuticle. **Smell** faint. **Taste** of hazelnuts. **Spores** olivaceous, 15–17 × 4·5–6 μ; fusoid. **Habitat** in late summer and autumn, sometimes in May, under frondose trees; occasionally under conifers; both on high and on low ground, in groups or alone. **Edible/Poisonous** edible; of good quality.

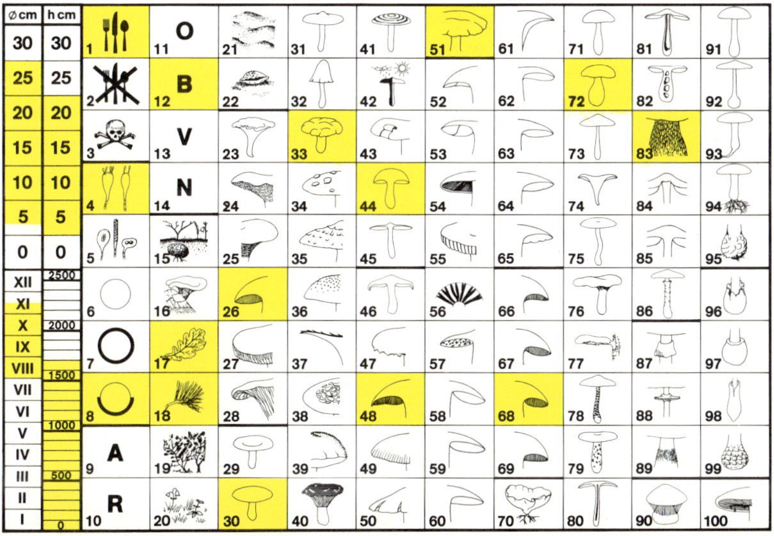

Cap hemispherical, blackish-brown, occasionally with darker and paler patches; later convex with wavy margin, somewhat faded, dry and velvety; with dark red crevices. **Tubes** 1·5–2·5cm long; at first tangential to the stipe, then free; white, later greenish. **Stipe** swollen, later almost cylindrical, thinner at the top; white or brownish, with pale reticulations; full. **Flesh** firm, white. **Smell** pleasant. **Taste** pleasant. **Spores** olivaceous-brown, 12–16 × 4–5 μ; fusoid. **Habitat** under frondose trees, both on high and low ground. **Edible/Poisonous** edible; of good quality.

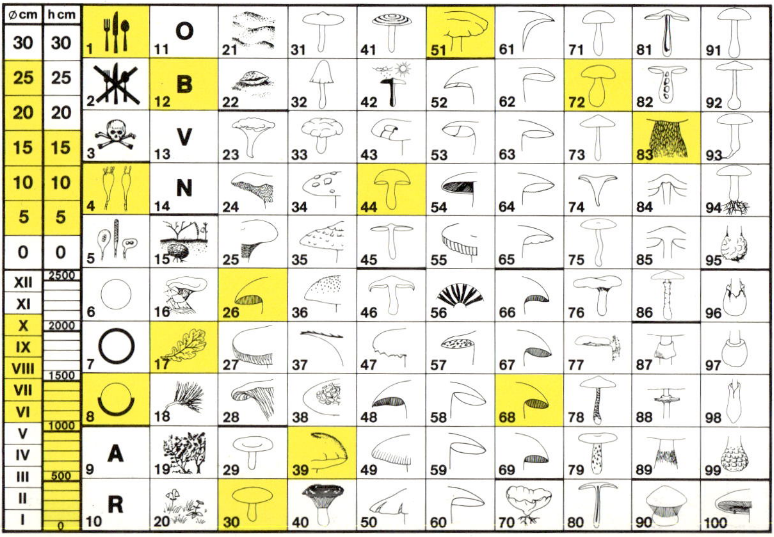

Lurid Bolete

Cap hemispherical, then less convex with involute margin; finally plane; dry and finely velvety; ochraceous or brown, occasionally tinged with olive-green; turning bluish when bruised, like the rest of this mushroom. **Tubes** 2–3·5cm long, adnexed at first, then free; yellow, then greenish; detachable in groups; the surface upon which they are set is red, turning blue when exposed to the air. **Pores** small, round, orange then red. **Stipe** variously shaped; yellow, covered in reticulations with oblong red mesh; full. **Flesh** yellowish in the cap, reddish in the foot. **Smell** pleasant. **Taste** pleasant. **Spores** olivaceous, 11–16 × 5–7·5 μ; fusoid. **Habitat** from spring to autumn in woods or on the edge of woods; in groups or alone; common. **Edible/Poisonous** poisonous when raw, edible if properly cooked.

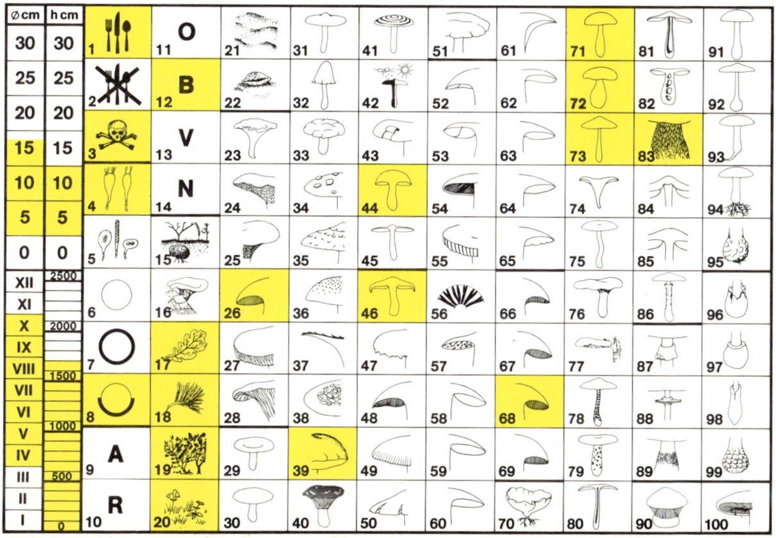

Boletus erythropus Persoon ex Fries

Red-stalked Bolete

Cap hemispherical, then expanded; from brown to reddish-brown or olivaceous-brown; velvety; margin long, involute. **Tubes** 1·3cm long, crowded, yellow then olivaceous, turning bluish when touched; free and implanted on a yellow surface. **Pores** small, round, orange then red. **Stipe** spotted with red on a yellow background. **Flesh** yellow, turning blue in the cap if exposed to the air; pinkish in the foot. **Smell** acid. **Taste** acid. **Spores** ochraceous-olivaceous, 12–17 × 5–7 μ; fusoid. **Habitat** from spring to autumn in woodland clearings. **Edible/Poisonous** poisonous if raw; edible once cooked.

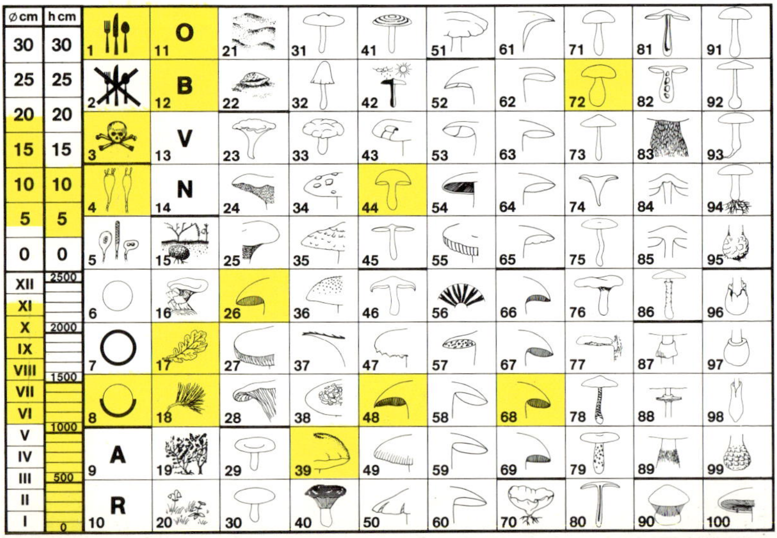

Brown Birch Bolete
Syn. *Leccinum scabrum* (Fr.); *Boletus scaber* (Fr.)
Cap with drooping cuticle, not easily peeled; damp, velvety, sometimes slightly rough; brown, occasionaly tinged with blackish, fading with age to ochraceous; very fleshy. **Tubes** 2–3cm long, from adnate to free, white or whitish, becoming ash-pink in the air and rusty-grey with age. **Pores** small; concolorous with the tubes. **Stipe** at first swollen, then slender and tapering toward the summit; ash-grey, with rows of darker scales; full. **Flesh** somewhat soft in the cap, fibrous in the stipe; white, later with greenish tinge. **Smell** faint. **Taste** mild. **Spores** pale brown, 13–20 × 5–6 μ; fusoid. **Habitat** under birches and poplars. **Edible/Poisonous** edible; of good quality.

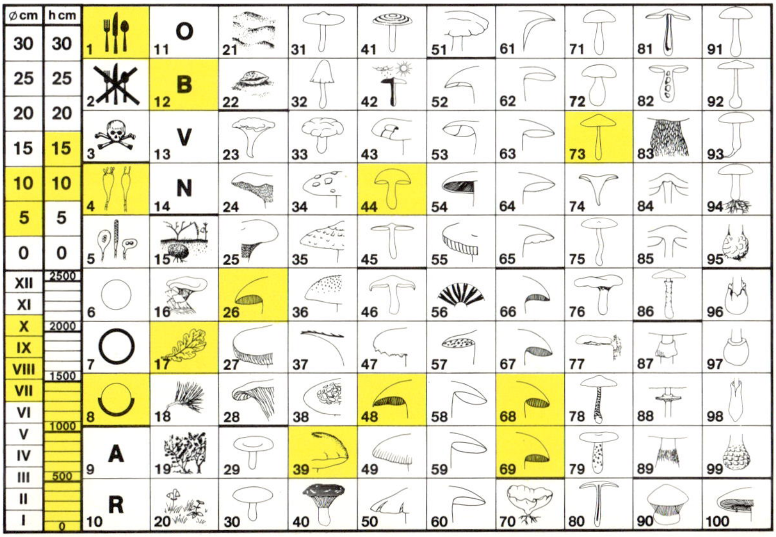

Granulated Bolete
Cap convex, with involute margin, then expanded, occasionally retaining a prominent apex; viscid, easily peeled; brown. **Tubes** 1–2cm long, adnate or almost free; yellowish, later tinged with greenish. **Pores** minute, angular; yellowish, then tinged with greenish; they exude droplets which leave behind a dark mark. **Stipe** whitish then brownish; near the cap it exudes droplets which leave behind a dark mark. **Flesh** soft and damp; yellowish. **Smell** pleasant. **Taste** pleasant. **Spores** olive-brown, 8–10 × 3–4 μ; fusoid. **Habitat** under conifers, particularly pines; common. **Edible/Poisonous** inedible if old; edible only when young.

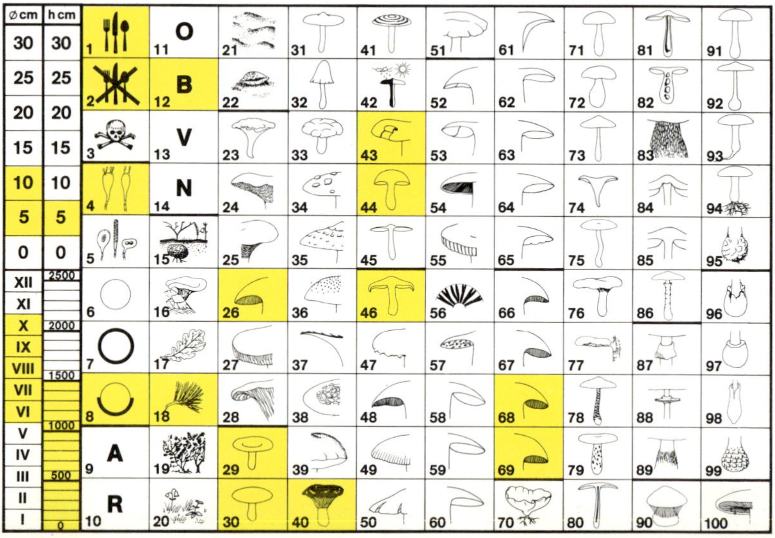

Bay Bolete

Cap hemispherical, then expanded and even plane; viscid in wet weather, shiny in dry weather; velvety, brown. **Tubes** 1–1·5cm long, from adnexed to adnate; whitish then greenish. **Pores** tiny, later slightly wider; white or greyish, then greenish; becoming blue when bruised. **Stipe** slightly swollen, then almost cylindrical; fibrillose, lighter than the cap; full. **Flesh** firm, but soon flaccid in the cap; fibrous in the stipe; whitish; brownish immediately underneath the cuticle, turning bluish in the air. **Smell** pleasant. **Taste** pleasant. **Spores** olive-brown, 12–16 × 4–5 μ; fusoid. **Habitat** in the woods in late summer and autumn; in groups. **Edible/Poisonous** edible; of good quality.

Mock Oyster or Hollow-stemmed Bolete
Cap with involute and festooned margin; umbonate or with prominent apex; woolly, sometimes scaly or hairy; brown or yellowish-brown. **Tubes** short, decurrent, yellow then olivaceous. **Pores** large, angular, regularly arranged, yellow then greenish-yellow. **Stipe** slightly lighter than the cap, fibrillose, with ring; hollow. **Flesh** damp, yellowish. **Smell** faint. **Taste** mild and sweetish. **Spores** yellow-olivaceous, 7·5–11 × 3·5–4·5 μ; fusoid. **Habitat** in late summer and autumn at the foot of larches; in groups. **Edible/Poisonous** edible.

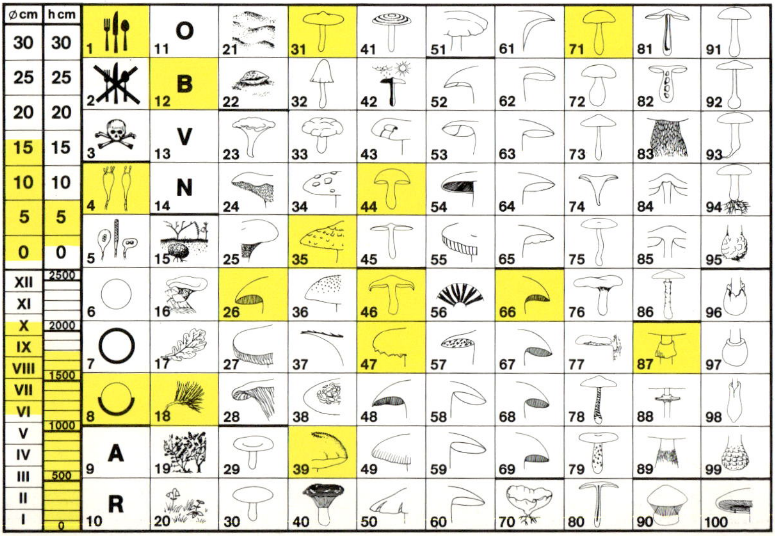

Polyporus pescaprae Persoon ex Fries

Cap gibbous; at first with involute margin, then unrolled and divided into two or three lobes; covered in fibrillose scales, regularly arranged and sooty-brown, which occasionally show the lighter cuticle. **Tubes** very short, decurrent; whitish. **Pores** wide, angular, white then ochraceous. **Stipe** eccentric, swollen, with a deep longitudinal furrow; whitish, shaded with ochraceous, scaly; full. **Flesh** firm but rather fragile; whitish, tinged with yellow. **Smell** pleasant. **Taste** pleasant. **Spores** white, 8–10 × 5–6·5 μ; elliptical. **Habitat** in the woods and in grass. **Edible/Poisonous** edible while young, even when raw.

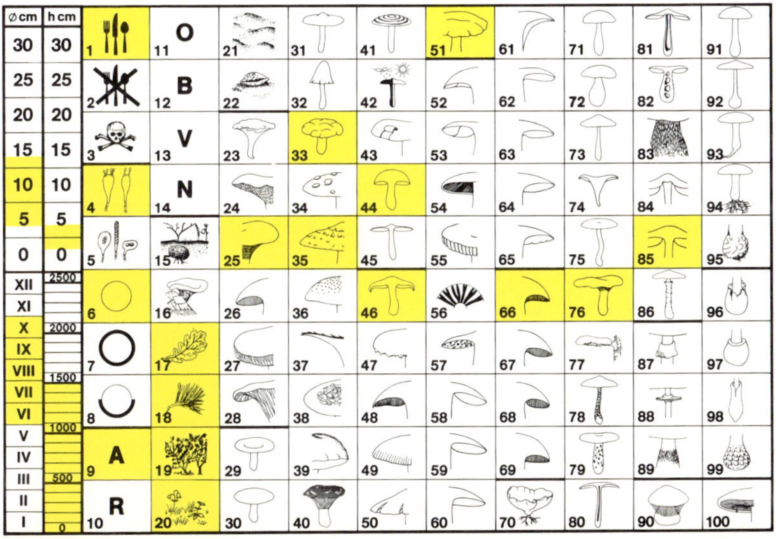

Imbricated Hydnum
Syn. *Hydnum imbricatum* (Linn ex Fr.)
Cap with involute margin, then expanded; often infundibuliform; brownish, with rough scales arranged concentrically, disclosing the whiter background. **Spines** decurrent, crowded; white, then ash-grey and finally yellowish; very fragile. **Stipe** often irregular, whitish, scaly, eventually brownish; full, sometimes hollow. **Flesh** firm, ash-grey. **Smell** pleasant. **Taste** bitterish. **Spores** brown, 6–7 × 5–6 μ; polygonal. **Habitat** under conifers, in tufts or groups. **Edible/Poisonous** not to be recommended when fresh, although not poisonous, because of its unpleasant taste even if properly cooked. Edible, and of good quality when dried.

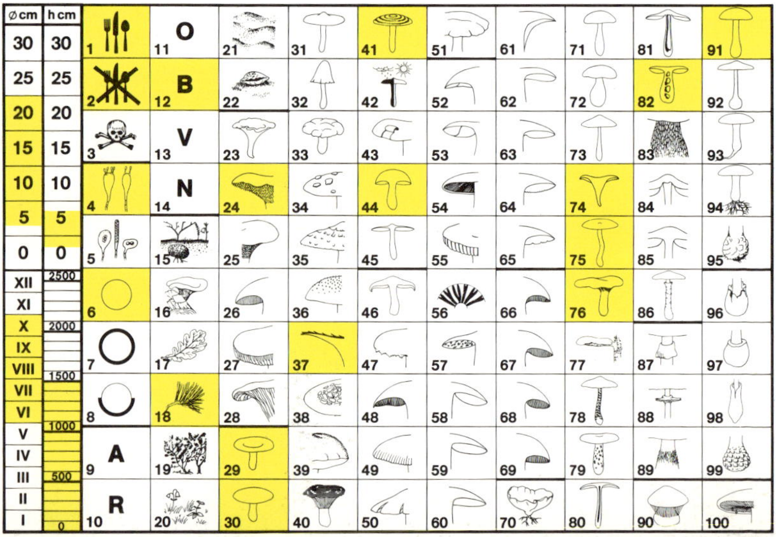

Cap irregularly globular, sometimes geminate, with corrugations resembling those of a brain, adhering to the stipe at various levels; from reddish-ochre to brownish. **Stipe** stumpy, with pitted surface, pruinose and whitish; hollow. **flesh** fragile, yellowish-white. **Smell** mild. **Taste** faint. **Spores** white, 25–25 × 12–15 μ; elliptical, elongated at the poles, with drop. **Habitat** in spring under conifers alone or in small groups; not common. **Edible/Poisonous** inedible when raw; poison not always destroyed during cooking.

Globular with concave or deeply furrowed base; black with brownish patches; covered with tiny warts more or less protruding. **Flesh** firm and whitish, then brownish with whitish marbling confluent at the base. **Smell** of creosote. **Taste** bitterish. **Spores** brown, 27–50 × 25–35 μ; elliptical-ovoid, spiky and reticulated. **Habitat** from early spring to winter in the Mediterranean areas; under oak trees. **Edible/Poisonous** edible in small quantities.

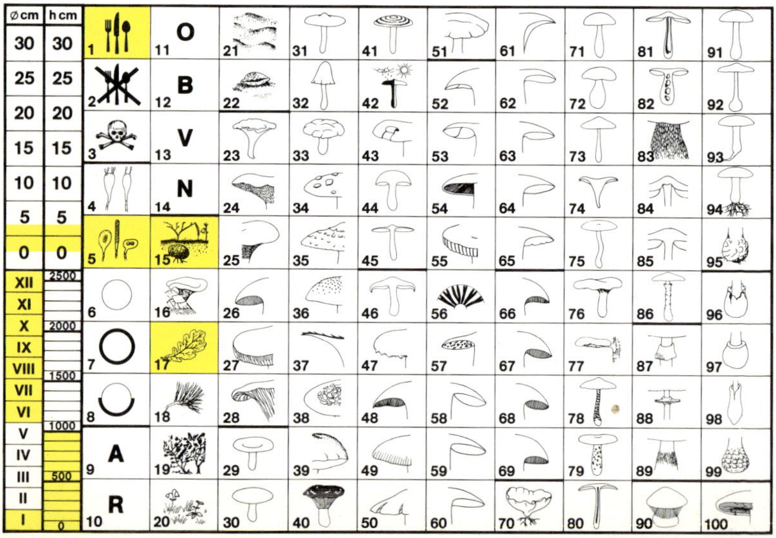

ETYMOLOGY OF SCIENTIFIC TERMS

(l. = Latin   g. = Greek)

acutesquamosa: l. *acutus* = sharp, pointed, l. *squamosus* = scaly; with sharp scales.
aegerita: g. *aigeiros* = poplar; of poplars.
aereus: l. = bronze-like; in colour or hardness.
aeruginosa: l. *aerugo* = verdigris; the colour of verdigris.
aggregatum: l. = joined together; in tufts.
AGROCYBE: g. *agros* = field, g. *kube* = head; field-head.
albonigra: l. *albus* = white, l. *niger* = black; both black and white.
AMANITA: g. *Amanos* = a mountain between Cilicia and Syria.
arenicola: l. *arena* = sand, l. *colore* = to live in; living in the sand (habitat).
ARMILLARIA: l. *armilla* = bracelet; having a ring.
arvensis: l. *arvum* = field; growing in ploughed fields.
atramentarius: l. *atramentum* = ink; referring to the colour of the spores.
atrotomentosus: l. *ater* = black, l. *tomentosus* = woolly; having a stipe covered in blackish wool.
aurantia: l. *malum aurantium* = orange; orange-coloured.
aurantiacus: orangey (same origin as above).

badius: l. = chestnut brown, reddish brown.
bitorquis: l. *bis* = twice, l. *torquis* = necklace; with two rings.
BOLETUS: g. *bolos* = sod; sod-like (in shape and/or colour).
bombycina: l. *bombyx* = silk worm; silky.
BOVISTA: puff-ball (German dialect *bofist*); due to the cloud of spores emitted at the slightest pressure.

caesarea: l. = worthy of the Caesars.
caligatum: l. *caliga* = boot; with the volva wrapped around the stipe.
CALVATIA: l. *calvus* = bald; with glabrous surface.
campestris: l. *campus* = field; growing in the fields.
cancellatus: l. *cancellum* = iron trellis, gate; reticulated.
CANTHARELLUS: g. *kantharos* = cup; small cup.
caperata: l. *capera* = wrinkle; wrinkled.
capnoides: g. *kapnodes* = smoky, smoked.
caprinus: l. = goat-like; so called because growing in pastures where goats are normally found.
castaneus: l. = chestnut-like; the colour of chestnuts.
cavipes: l. *cavus* = hollow, l. *pes* = foot; with hollow foot or stipe.
cerussata: l. *cerussa* = ceruse; white, silvery-white.
cervinus: l. = the colour of deer.
cibarius: l. = edible.
cinnabarinus: g. *kinnabarinos* = red like cinnabar, vermilion.
citrina: l. *citrus* = lemon; lemon yellow.
CLATHRUS: l. *clathri* = iron bars, grating; reticulated.
clavatum: l. *clava* = club; club-shaped.
CLAVARIA: same as above.
clavipes: l. *clavus* = nail, l. *pes* = foot; with nail-like, or swollen, foot.

CLITOCYBE: g. *klitus* = slope, g. *kube* = head; with sloping head (usually when the gills are deeply decurrent).
CLITOPILUS: g. *klitus* = slope and g. *pileos* = cap; same as above.
clypeatum: l. *clypeus* = shield (round); shaped like a shield.
coccinea: l. *coccinus* = scarlet.
colossus: l. = giant.
columbetta: l. *columba* = dove; small dove (because white as a dove).
comatus: l. *coma* = hair; refers to the scales on the cap.
confluens: l. = confluent.
controversus: l. = turning upwards.
COPRINUS: g. *kopros* = dung; refers to the habitat.
cornucopioides: l. *cornucopia* and g. *eidos* = image; similar to a cornucopia.
CORTINARIUS: having a cortina.
CORTINELLUS: diminutive of the above.
CRATERELLUS: l. *crater* = cup; small cup.
crispa: l. *crispum* = curly; wavy, curly.
crustuliniforme: l. *crustulina* = thin crust (of bread); the appearance of bread crust.
cyanescens: g. *kuanos* = blue; becoming blue.
cyanoxantha: g. *kuanos* = blue, g. *xanthos* = yellow; of these, or intermediate, colours.
cyathiformis: g. *kuathos* = cup; in the shape of a cup.

delica: l. = without milk.
deliciosa: l. = delicious.
duriusculus: l. *durus* = hard; rather hard.

edodes: g. *edodes* = food; edible.
edulis: l. = edible.
elegans: l. = elegant.
emetica: l. = causing vomiting.
equestre: l. *eques* = knight; reserved to the nobility.
ENTOLOMA: g. *entos* = inwards, g. *loma* = margin; with involute margin.
erythropus: g. *eruthros* = red, g. *pous* = foot; with a red foot.
esculenta: l. edible.
excoriata: l. = peeled, excoriated.

fasciculare: l. *fasciculus* = small bundle or bunch; growing in small tufts.
fastigiata: l. *fastigium* = pointed roof; ending in a point.
fellea: l. = as bitter as bile.
ferrii: of G. Ferri.
fistulous: l. = resembling a pipe or tube.
flava: l. = yellow.
floccose: l. = covered with tufts.
foetens: l. = stinking, foetid.
frondosus: l. = frondose, with fronds, leafy.
fugacious: l. = falling or fading early.
fuliginosus: l. = sooty.
fuscous: l. = dark, dusty.
fusipes: l. *fusus* = spindle, l. *pes* = foot; with spindle-like foot.
fusoid: l. = spindle-like

gemmata: l. = covered in jewels; covered in jewel-like warts.
georgii: of St George (St George's Day is 23 April, the approximate date when it first appears).
GANODERMA: g. *ganos* = shine, g. *derma* = skin; with shiny surface.
gibbous: l. = convex, rounded.

gigantea: g. *gigantios* = gigantic.
glabrous: l. = free from hair, smooth.
gloiocephala: g. *gloios* = gluten, g. *kefale* = head; with glutinous head.
GOMPHIDIUS: g. *gomfos* = nail, g. *eidos* = image; nail-like.
grammopodia: g. *gramme* = mark, sign, g. *podion* = small foot; with marked foot.
granulatus: l. = granular.
GUEPINIA: of Guepin.
GYROCEPHALUS: g. *guros* = round, g. *kefale* = head; with a round head.

hadriani: of Hadrian.
HEBELOMA: g. *hebe* = down, g. *loma* = margin; with downy margin.
HELVELLA: l. = general term for aromatic herbs.
helvelloides: g. *eidos* = image, similarity; similar to Helvella.
helveola: l. *helvus* = yellow; yellowish.
hepatica: l. *hepas* = liver; the colour of liver.
hortensis: l. = of gardens and kitchen gardens.
HYGROPHORUS: g. *hugros* = humid, g. *fero* = I bring; having a damp cap.

imbricatum: l. = covered in roof-tiles; a scaly cap.
infundibuliformis: l. *infundibulum* = funnel; funnel-shaped.
INOCYBE: g. *is* = plant fibre, g. *kube* = head; with fibrous cap.
integra: l. = integral, whole.
involutus: l. = inward-turning.

LACTARIUS: l. *lac* = milk; containing milk.
LENTINELLUS: small Lentinus.
LENTINUS: l. *lentus* = strong, tough; so called because of its tough flesh.
lepida: l. = pleasant, likeable.
LEPIOTA: g. *lepis* = scale, g. *ous* = ear; having ear-shaped scales on the cap.
LEUCOCEPHALUS: g. *leukos* = white, g. *kephalo* = head; with a white cap.
leucophaeus: g. *leukos* = white, g. *faios* = dark; white and greyish.
lilacina: Persian *lilac* = pinkish-violet.
lividum: l. = livid, leaden.
lucidum: l. = shiny.
luridus: l. = lurid, squalid.
lutescens: l. *luteus* = yellow; tending to yellow.
LYCOPERDON: g. *lukos* = wolf, g. *perdomai* = bladder; expelling a mass of dark spores at the slightest pressure.
LYSURUS: g. *luo* = to untie, g. *oura* = tail; alluding to the terminal tentacles.

macrocephalus: g. *makros* = large, g. *kefale* = head; with a large head.
maculata: l. = spotted, patchy.
MARASMIUS: g. *marasmos* = loss of weight; these mushrooms are often stringy and thin.
marzuolus: l. = of the month of March.
MELANOLEUCA: g. *melas* = black, g. *leukos* = white; black and white.
mellea: l. *mel* = honey; because of the colour of the cap.
meridionale: l. *meridies* = midday, south; of a southerly habitat.
mesentericum: g. *mesenterion* = intestinal membrane; similar to a mesenterium.
mokusin: a region of China.
MORCHELLA: from the German *morchel*, morel, edible fungus.
mucida: l. = viscous.
MUCIDULA: diminutive of the above.
muscaria: l. *musca* = fly; it was used to kill flies.
mutabilis: l. = changeable.
MYCENA: g. *mukes* = mushroom.

naucina: l. *naucum* = walnut shell.
nebrodensis: from Nebrodi, a range of mountains in Sicily.
nebularis: l. *nebula* = cloud; because of the colour of its cap.
NEMATOLOMA: g. *nema* = thread, g. *loma* = edge, margin; with filamentose margin.
NEVROPHYLLUM: g. *neuron* = nerve, g. *fullon* = leaf, gill; having veins (or nerves) instead of gills.
nigricans: l. = tending to black.
NOLANEA: l. *nola* = bell; for the shape of the cap.

odora: l. = sweet-scented.
olearia: l. = pertaining to olive trees.
olivacea: l. = the colour of olives.
oreades: the name given to wood nymphs.
orellanus: g. *oros* = mountain; of the mountains.
ostreatus: l. *ostrea* = oyster; similar to a tuft of oysters.
ovinus: l. *ovis* = sheep; sought after by sheep.
ovoidea: l. *ovum* = egg; ovoid, due to its initial shape.

pantherina: l. *panthera* = panther; due to its colour.
pascua: l. = of the pastures.
patouillardii: of N. Patouillard.
PAXILLUS: l. = small stick.
pes caprae: l. = goat's foot; due to its overall shape.
PEZIZA: g. *pezis* = mushroom, usually small.
phalloides: g. = similar to a phallus.
piperatus: l. *piper* = pepper; peppery.
PLEUROTUS: g. *pleuron* = side, g. *ous* = ear; with the cap on one side, ie with eccentric stipe.
plumbea: l. = the colour of lead.
PLUTEUS: l. = shelf; growing like a shelf.
poetarum: l. = of the poets.
POLYPORUS: g. *polus* = much, g. *poros* = pore; with many pores.
populinus: l. *populus* = poplar; of the poplars.
portentosus: l. = portentous, prodigious, extraordinary.
praecox: l. = premature.
praestans: l. = handsome.
pratense: l. = of meadows.
pratensis: see above.
procera: l. = grown, tall.
pruinose: l. = covered with a fine white powder like hoar frost.
prunulus: l. *prunus* = plum.
PSALLIOTA: g. *psallion* = curb, chain, g. *ous* = ear; with fringed cap.
puniceus: l. = purple.
purpureus: l. = purple.

radicata: l. = rooted; due to the root-like extension of the stipe.
repandum: l. = turned upwards.
rhacodes: g. = shaggy; because of the scales on the cap.
RODOPAXILLUS: g. *rhodon* = pink, rose, l. *paxillus* = stick; pinkish stick.
rotunda: l. = round.
ROZITES: after E. Roze.
rubescens: l. = becoming red.
rufus: l. = fuscous red or golden red.
RUSSULA: *russus* = red; reddening, although many Russulae are of another colour.
rutilans: l. = rutilant, glowing.

saponaceum: l. *sapo* = soap.
SARCODON: g. *sarx* = flesh, g. *odous* = tooth; with fleshy teeth.
SARCOSCYPHA: g. *sarx* = flesh, g. *skufos* = cup; fleshy cup.
satanas: l. = Satan; because of its poison.
SCLERODERMA: g. *skleros* = hard, g. *derma* = skin; having a hard skin.
scrobiculatus: l. *scrobiculus* = small erosion; with small erosions on the stipe.
sepium: l. = of the hedges.
solitaria: l. = solitary.
SPARASSIS: g. *sparaxis* = lacerated.
squarrosa: l. = covered in pimples.
squamosus: l. = scaly.
STROPHARIA: l. *strophium* = pectoral band; so called because of its ring.
subbotrytis: l. *sub* = almost; similar to the Botrytis.
sulcate: l. = marked with parallel grooves.
sulphureum: l. = sulphureous.

tabescens: l. *tabescere* = to dissolve, to liquify (oneself); liquifying.
terreum: l. = earth-like; earth-brown.
tigrinum: l. = the colour of tigers.
torminosus: l. *tormina* = colic; provoking colics.
traganus: g. *tragos* = ram; due to its smell.
TRICHOLOMA: g. *thrix* = hair, g. *loma* = margin; with hairy margin, however sparse.
TUBER: l. = tuber; truffle.
turbinate: l. = inversely conical, narrow at bottom, wide at top.

umbellatus: l. *umbella* = umbrella; composed of many small umbrellas.

vaginata: l. *vagina* = sheath; due to the shape of the volva.
vellereus: l. *vellus* = vellum, wool; woolly.
velutipes: l. *velum* = veil, l. *pes* = foot; with velvety foot.
verna: l. *ver.* = spring; growing in spring.
vesca: l. *vescus* = edible.
violaceus: l. = violaceus.
virescens: l. = tending to or becoming green.
virosa: l. *virus* = poison; poisonous.
viscidus: l. = viscid.
volemus: l. *vola* = palm (of the hand); producing enough milk to fill a hand.
VOLVARIA: l. *volva* = volva; with a volva.

xanthoderma: g. *xanthos* = yellow, g. *derma* = skin; with yellow skin (when rubbed).

SYSTEMATIC INDEX OF LATIN NAMES

Numbers given refer to colour plates

ALPHABETICAL INDEX OF LATIN NAMES

Numbers given refer to colour plates

ALPHABETICAL INDEX OF VERNACULAR NAMES

Numbers given refer to colour plates

WILD HERBS: A FIELD GUIDE
Jacques de Sloover and Martine Goosens

Whether used as a practical identification guide in the field, or for armchair browsing, this book offers a great deal of information succinctly presented. The herbs in the 144 stunning colour plates are grouped by colour. To identify a plant, simply open the guide at the pages bordered by the colour corresponding to that of the flower and you will soon find the plant itself. This ingenious system is time-saving, and the close proximity of illustrations of similarly-coloured flowers helps to avoid misidentification.

For the purposes of this book, a herb is defined as a useful plant, one which is used to cure, to feed, to flavour dishes, to dye wool, or for any other specific aim. Pictograms presented alongside each colour plate summarize other properties – aromatic, medicinal and culinary, which parts are efficacious, when the herb is at its prime, where it grows, when it flowers. This at-a-glance information is supplemented by useful appendices, a glossary and notes on further reading.

210×120mm (8¼×4¾in) 144 colour plates, 144 line illustrations and keys

MINERALS & GEMSTONES: AN IDENTIFICATION GUIDE
Guiseppe Brocardo

A completely new way to approach the mineral world is here offered to the experienced collector and the beginner. The 156 splendid colour photographs are accompanied by a brief description and a pictographic table which provides, through easily recognisable symbols, the available information necessary for identification and classification. The key to the symbols is printed on a bookmark. The margins of the pages are coloured to reflect those of the minerals themselves and aid quick identification.

The colour plates are preceded by extensive information on how to recognise and collect minerals, their origin and formation, their structure and properties, the classification, and how to prepare them for preservation. A glossary, bibliography and indexes complete the volume and add to its value as an indispensable guide for all collectors.

Guiseppe Brocardo was born in 1916 and has a degree in natural sciences. For the past 35 years he has been teaching in Turin, and is director of the Museo di Storia Naturale there. He has personally photographed the museum's collection to illustrate this book, and is a frequent contributor to magazines and periodicals on mineralogy.

210×120mm (8¼×4¾in) 156 colour plates, 186 line illustrations and keys